Condensed Matter Physics
The Theodore D. Holstein Symposium

Theodore D. Holstein

Raymond L. Orbach, Editor

Condensed Matter Physics

The Theodore D. Holstein Symposium

With 71 Illustrations

Springer-Verlag
New York Berlin Heidelberg
London Paris Tokyo

Raymond L. Orbach
College of Letters and Sciences
University of California, Los Angeles
Los Angeles, CA 90024
U.S.A.

Library of Congress Cataloging in Publication Data
Theodore D. Holstein Symposium (1986: University
of California, Los Angeles)
Condensed matter physics.
1. Condensed matter—Congresses. 2. Polarons—
Congresses. 3. Electron-phonon interactions—Congresses.
4. Holstein, Theodore David, 1915–1985. I. Orbach, R.
II. Holstein, Theodore David, 1915–1985. III.Title.
QC173.4.C65T48 1986 530.4′1 87-9474

Printed and bound by R.R. Donnelley and Sons, Harrisonburg, Virginia.
Printed in the United States of America.

9 8 7 6 5 4 3 2 1

ISBN 0-387-96528-9 Springer-Verlag New York Berlin Heidelberg
ISBN 3-540-96528-9 Springer-Verlag Berlin Heidelberg New York

Preface

Theodore David Holstein died May 8, 1985, at the age of 69. His research career covered 46 years. His contributions have been seminal throughout this period, beginning with his first papers with H. Primakoff in 1939 and extending to the year of his death.

"Ted" earned his Ph.D. in physics from New York University in 1940, after earning his Master's degree from Columbia University in 1936 and his B.S. from N.Y.U. in 1935. After receiving recognition while he was a graduate student for his contributions to the atomic theory of magnetism, he participated in the development of radar at the Westinghouse Research Laboratories, where he was a research physicist from 1941 to 1959. He taught on the faculty of the University of Pittsburgh from 1959 to 1965. He joined the Physics Department of the University of California, Los Angeles, where he remained until his death. Ted is survived by his wife Beverlee, his daughter Lonna Smith, his son Stuart, and his grandson Andy Smith.

Ted received many prestigious awards and honors, including membership in the National Academy of Sciences and the American Academy of Arts and Sciences. He received a von Humboldt fellowship for research at the University of Regensburg in the Federal Republic of Germany.

The Theodore D. Holstein Symposium was held in Ted's memory on the campus of the University of California, Los Angeles, and at its Conference Center in Malibu, March 28–29, 1986. There were 73 attendees, spanning the research community from graduate students to Nobel Laureates. All were invited to contribute either original research articles or memorabilia to this volume. We are grateful to all those who have contributed to the Symposium, especially the financial support of the Westinghouse Research Laboratories and the University of California, Los Angeles.

Those of us who had the privilege of working with him know that we were all Ted's "students." We feel a profound emptiness which will not be filled. But Ted's legacy of integrity and dedication will long remain. His scientific papers are classics: incisive, complete, and richly detailed. Many of them laid the foundation for all future work in their area and related areas. A glance at his bibliography makes clear the contributions he has made to physics, and the reason so many of us are indebted to his scientific leadership.

Ted's thesis work with H. Primakoff in 1940 gave birth to the Holstein–Primakoff transformation which still forms the basis for treatments of spin-waves in ordered magnetic materials. His work on the imprisonment of resonant radiation in the late 1940s established a new field and is a model of clear, insightful analysis. It has been found useful not only for gases, but also for the solid state; not only for photons, but also for phonons. Ted's short article on the optical and infrared volume absorptivity of metals in the early 1950s has since been found to be fundamental for the analysis of the infrared absorption of normal metals and superconductors. Typically, the full work can be found in a Westinghouse unpublished report. Nevertheless, his short published article has had great impact.

This work was followed by half a decade of silence, to be broken by a veritable torrent of profound contributions. His calculation of the collision-drag effect laid the microscopic foundation for the beautiful macroscopic treatment of Pippard. His powerful and singularly successful series of papers on the polaron appeared the same year. Everyone who works on the polaron problem must make reference to these seminal papers. Again, at nearly the same time, he began a series of papers connected with the motion of a Bloch electron in a magnetic field. This led to the analysis of the Hall effect for impurity (hopping) conduction; high-frequency cyclotron resonance in an electron-phonon gas (with H. Scher); Hall mobility of the small polaron (with Lionel Friedman); the side-jump mechanism for the ferromagnetic Hall effect (with S.K. Lyo); the adiabatic theory of the Hall effect (with David Emin); the ferromagnetic Hall effect in the electron-phonon gas (with S.K. Lyo); and the ac ferromagnetic Hall effect (with S.K. Lyo). Taken together, these papers have had an impact on the entirety of the condensed matter field.

One of Ted's most elegant works is his study of the transport properties of the electron-phonon gas, some 149 pages in length, which appeared in the *Annals of Physics* in 1964. This monumental work, and especially its appendices, laid the foundation for the Boltzmann equation. It is a remarkably clear and complete work, and is a very effective textbook in its own right. It clearly displays his complete mastery of diagrammatics, and more important, his physical understanding.

In the mid-1970s, Ted clarified the microscopic basis of the Forster–Dexter theory of excitation transport in a series of papers (with S.K. Lyo and R. Orbach). This work forms the basis for the analysis of fluorescent line narrowing spectroscopy. His interest in spectroscopy led to the first successful microscopic theory for cooperative optical absorption (with S. Alexander and R. Orbach).

In the late 1970s, Ted began an extraordinary series of papers using a "simple" model to provide a quantitative and full expression for a chemical-rate theory of polaronic hopping. He then began a monumental series of papers exploring higher-order effects in hopping-type transitions of small polarons. This led to work completed just before his death on the motion of solitons in a dynamic lattice (with B. Schutler). He approached the problem of phonon–soliton interaction in the same fashion which marked so much of his work. He introduced a "simple" model which exhibited all of the important physics and then proceeded to analyze it in extraordinary depth. The full physical content of the problem emerged quantitatively.

This listing does not do justice to Ted's contribution to science. One must understand that the papers themselves are models of clarity. They bring order to subjects which previously had not been clear as to their microscopic basis. His work provided an inspiration to his generation, and to its students. It will remain a living monument; subsequent generations of scientists will turn to it for understanding and inspiration.

Ted's memory will inspire us all. He was one of the leading theoretical physicists of his generation. He was the quintessential physicist, who did not tolerate sloppy thought in anyone's work, especially his own. Ted was uncompromising in his search for understanding in physics.

It is for all these reasons that this symposium was organized. Those of us who were privileged to know and work with Ted hope that our contributions will acquaint others with his insight and accomplishments.

This volume is dedicated to his memory. We all miss him greatly.

Raymond L. Orbach
April, 1987

Contents

Contributors

PHILIP W. ANDERSON
Department of Physics
Jadwin Hall, P.O. Box 708
Princeton University
Princeton, NJ 08544

M. YA AZBEL
School of Physics and Astronomy
Tel Aviv University
Tel Aviv, Israel

P. BAK
Brookhaven National Laboratory
20 Pennsylvania St.
Upton, NY 11973

P.M. CHAIKIN
Department of Physics
University of Pennsylvania
Philadelphia, PA 19104

B.H. CHOI
Department of Physics
University of California, Riverside
Riverside, CA 92521

MARVIN L. COHEN
Department of Physics
University of California
and
Material and Molecular Research
Division
Lawrence Berkeley Laboratory
Berkeley, CA 94720

DAVID EMIN
Sandia National Laboratories
Division 3532
V.P. 1000, P.O. Box 5800
Albuquerque, NM 87185

O. ENTIN-WOHLMAN
Department of Physics
University of California, Los Angeles
Los Angeles, CA 90024
and
School of Physics and Astronomy
Tel Aviv University
Tel Aviv, Israel

A. THEODORE FORRESTER
(Deceased)
Department of Physics
University of California, Los Angeles
Los Angeles, CA 90024

LIONEL FRIEDMAN
Department of Electrical Engineering
Worcester Polytechnic Institute
Worcester MA 01609

T.H. Geballe
Stanford University
Stanford, CA 94305
and
Bell Communications Research
Murray Hill, NJ 07974

Edward Gerjuoy
Department of Physics and Astronomy
University of Pittsburgh
Pittsburgh, PA 15260

V. Jaccarino
Department of Physics
University of California, Santa Barbara
Santa Barbara, CA 93106

Walter Kohn
Department of Physics
University of California, Santa Barbara
Santa Barbara, CA 93106

M. Lagos
Facultad de Fisica
Universidad Catolica
Santiago, Chile

Moises Levy
Department of Physics
University of Wisconsin, Milwaukee
P.O. Box 413
Milwaukee, WI 53201

N.L. Liu
Department of Physics
University of California, Riverside
Riverside, CA 92521

S.K. Lyo
Sandia National Laboratories
Division 3532
V.P. 1000, P.O. Box 5800
Albuquerque, NM 87185

D.J. Scalapino
Department of Physics
University of California, Santa Barbara
Santa Barbara, CA 93106

Susan C. Schneider
Department of Electrical Engineering and Computer Science
Marquette University
Milwaukee, WI 53233

J. Robert Schrieffer
Institute of Theoretical Physics
University of California, Santa Barbara
Santa Barbara, CA 93106

Ivan K. Schuller
Argonne National Laboratory
9700 S. Cass Ave.
Argonne, IL 60439

Bernd Schüttler
Department of Physics
University of California, Santa Barbara
Santa Barbara, CA 93106

X. Shen
Department of Physics
University of California, Riverside
Riverside, CA 92521

D. Shoenberg
Department of Physics
University of Cambridge
Madingly Road
Cambridge CB3 OHE, England

Leonid A. Turkevich
Standard Oil
Corporate Research
4440 Warrensville Center Road
Cleveland, OH 44128

LAWRENCE A. VREDEVOE
Saint Johns Hospital
1328 22nd St.
Santa Monica, CA 90404

Y. YAFET
AT&T Bell Laboratories
600 Mountain Ave.
Murray Hill, NJ 07974

Magnetic Field Induced Transitions in Organic Conductors and Gaps in the Rings of Saturn

P. M. Chaikin, M. Ya Azbel, and P. Bak

Abstract

Quasi-two dimensional organic conductors in a high magnetic field exhibit a series of remarkable phase transitions from normal metals to Spin Density Wave semi-metals. Recent experiments on $(TMTSF)_2ClO_4$ are presented. In attempting a theoretical treatment of these field induced transitions we were led an investigation of the effects of two widely different periodic potentials, a large, slowly varying potential from the open orbits and the magnetic field, and a small, rapidly varying potential from the SDW distortion. Similar equations have been proposed as an explanation of the anomalously large size of gaps in the rings of Saturn and in particular for the Cassini division. The novel spectrum associated with these incommensurate potentials leads to a series of bands and gaps, and in the limit of a large ratio of the periods, to an effective gap with the size associated with the slow period, but at the energy or frequency associated with the fast period. This confirms a conjecture due to Avron and Simon and may explain the size of the Cassini Division.

I. Introduction

Along with the brilliant papers that characterize the work of Ted Holstein, there are some contributions which have not been as widely cited. The work which is discussed in this lecture derives largely from one such paper, [1] written about three years ago with two of the present authors. Somewhat before, there had been an experimental report by Kwak et al [2] of magnetoresistance oscillations in the organic conductor $(TMTSF)_2PF_6$. The observation of these oscillations was quite surprising. Up until that time the material and its close relatives

($(TMTSF)_2X$ X= ClO_4, PF_6,ReO_4 etc.) had been described as electronically one-dimensional, and certainly as having nothing but open orbits at their Fermi surface. Magnetic oscillations have always been associated with the quantization required for closed orbits in two or more dimensions and the formation of Landau levels. The question that Ted and his collaborators addressed was whether it is possible to have magnetic oscillations for open orbits in two dimensions.

The basic idea was that in the presence of a perpendicular magnetic field an electron traveling on a two dimensional open orbit executes periodic motion in k space, returning to the same k state with a well defined frequency $\omega_c' \sim (v_{fx}H)(2\pi/b)$ as illustrated in Fig. 1a. In three dimensions an infinitessimal tilt of the open orbit would result in a different transit time through successive Brillouin zones. Since periodic motion is quantized one might expect that the energy $\hbar\omega_c'$ comes into the problem. In order to look for such an effect a model bandstructure which allowed open orbits was taken with a free electron form along x and tight binding (bandwidth $4t_b$) along y.

$$E(k_x,k_y) = (\hbar k_x)^2/2m + 2t_b\cos(k_y b) \qquad (1)$$

Taking the more conventional tight binding form along both directions makes a rigorous treatment more difficult, since it leads to the complex Azbel-Hofstadter [3,4] spectrum which we will encounter later. The magnetic field is then included with the usual Landau - Peierl's substitution, k -> i(grad)-eA/hc, and a Landau guage A=(0,Hx), leaving a differential equation to be solved for the eigenstates and energies. The variable y appears only as a derivative and therefore the wavefunction can be factored into a plane wave along y times a function of x, $Y(x,y)=\exp(ik_y y)Y(x)$. The equation then takes the form:

$$-(\hbar^2/2m)\partial^2 Y(x)/\partial x^2 + 2t_b\cos(k_y b-eHbx/hc)Y(x) = EY(x) \qquad (2)$$

and since k_y occcurs only in the argument of the cosine it mearly serves to shift the origin of x. The equation reduces to that of a one dimensional electron in a cosine potential, which from the Floquet-Bloch theorem has solutions with a continuous variable k_x (i.e. $E(k_x)$). Thus in the presence of a magnetic field the two dimensional dispersion relation in Eq.1 has been replaced by a one dimensional dispersion relation.

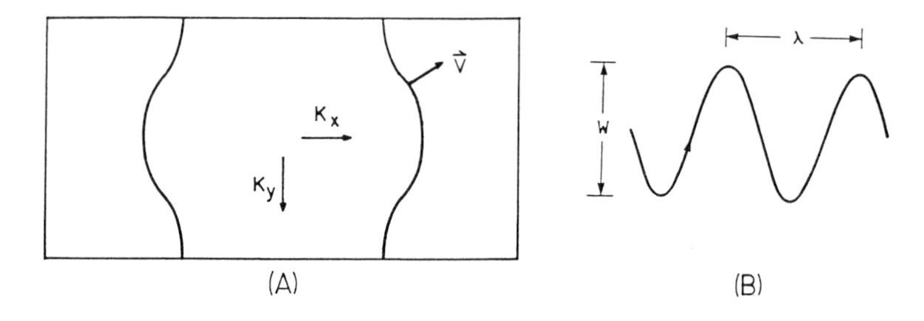

Figure 1. A) Open orbit Fermi surface extending from π/b to $-\pi/b$ from top to bottom along k_y. V is velocity which changes as the electron moves along the Fermi surface. B)Real space motion is extended along x and periodic along y.

The physics behind this unusual result, the one dimensionalization in the presence of a magnetic field, can be seen in a number of ways. In the absence of scattering and at zero temeprature an electron on the Fermi sureface traverses the Brillouin zone from left to right in Fig. 1a and therefore experiences all of the k_y states. Its energy cannot depend on k_y and only depends on the 'average ' value of k_x for the orbit. The real space motion is shown in Fig. 1b. The velocity of an electron is perpendicular to the Fermi surface as shown in Fig. 1a. As the electron moves across the open orbit Fermi surface in the presence of the magnetic field the velocity component along k_y changes sign whereas the component along k_x remains in the smae direction. In the absence of a magnetic field an electron on the Fermi surface can explore all of space in the directions x and y. In a magnetic field it has restricted motion in y ($w=\{v/\omega\}=4t_b b/\hbar\omega_c$) and infinite motion in x. It is in this sense that the electron motion is one dimensional. Note that the period along x as well as the width of the y excursions both are inversely proportional to the applied field.

The results of this calculation were quite discouraging. Eq. 2 is Mathieu's equation and the assymptotic solutions are well known: for E less than $2t_b$ there are large gaps in the spectrum corresponding to closed orbits, for $E > 2t_b$ (the case of interest, $E_f \sim 8t_b$) the gaps become exponentially small. The energies (ω_c', E_f, t_b) for the organic salts were such that the cross over between these regions was quite abrupt. Magneto-oscillations can be seen for open orbits, but not for the materials of interest. The paper also pointed out the criteria for the system to be regarded as two dimensional, the bandwidth in the third direction should be less than the Landau level splitting or the gaps in the two dimensional spectrum. This two-three dimensional crossover was suggested as the cause of the threshold field (Kwak field) giving the onset of the oscillations in the Kwak experiments.

II. Experiments

The experimental situation became significantly clearer with the measurement of the Hall coefficient by two separate groups [5,6]. The Hall resistance for $(TMTSF)_2ClO_4$ is shown in Fig. 2. Above the Kwak field the Hall resistance increases by ~ two orders of magnetude indicating that the carrier concentration has dropped by about that ratio. The Hall resistance then shows a series of plateaus reminiscent of the quantum Hall effect. However, the fact that there is only approximate periodicity in 1/H, coupled with the observation that the values of the resistances at the plateaus is highly temperature dependent led both groups to discount a simple quantum Hall effect interpretation. Instead the conclusion was that the material was undergoing a series of transitions to SDW semi-metal phases. At a later time, after some of the theories described below were proposed, experiments confirmed that the Hall steps corresponded to thermodynamic transitions [7,8].

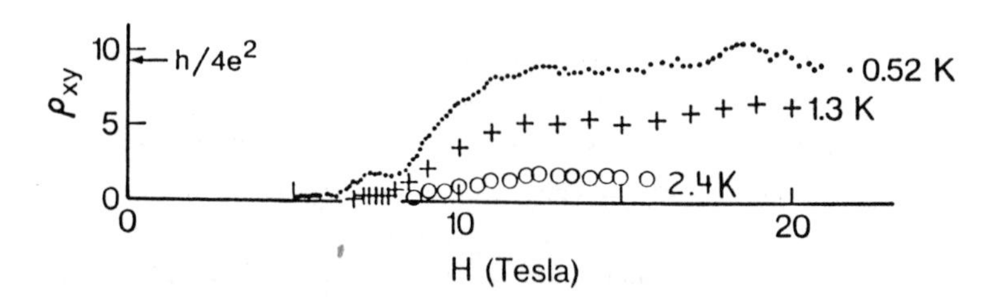

Figure 2. Hall resistance as a function of magnetic field for $(TMTSF)_2ClO_4$ at several temperatures.

One of the confirming experiments was the magnetization measurement [7]. The samples of the Bechgaard salts are high quality needle-like crystals with small mass (<1milligram) making conventional susceptiblity measurements difficult, especially in high fields and at low temperatures. A very simple modification of conventional transport apparatuses made the experiment possible. For resistance measurements a sample is mounted with four thin wires attached as voltage and current leads cantelevered from a sample holder. If a metal plate is placed below the sample, then the capacitance between the sample and the plate can be measured and the sample displacement monitored sensitively. A magnetic field will exert a force (MdH/dz) and a torque MxH on the sample. The four leads act as very weak (and adjustable) springs providing a restoring force. They can be calibrated for instance by turning the sample holder upside down and using the known gravitational force. The force and torque components can be separated by moving the sample to different positions in the field profile to get different gradients, or by using two capacitor plates at the opposite ends of the sample to directly observe torques.

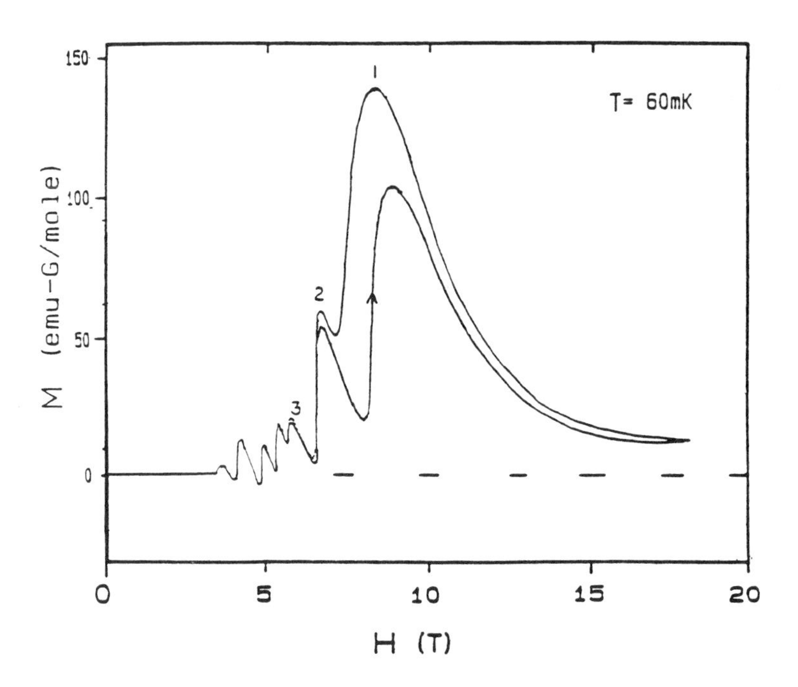

Figure 3. Magnetization of $(TMTSF)_2ClO_4$ as a function of magnetic field.

The magnetization of $(TMTSF)_2ClO_4$ at 60 milliK is shown in Fig. 3.

The presence of discontinuous jumps in the magnetization at finite temperatures show that the transitions are first order. The magnetic version of the Clausius - Clapeyron equation, $\Delta S=-\Delta M(dH/dT)$ says that if ΔM is finite, a jump, and the field at which these jumps is observed is temperature dependent, then there is a jump in the entropy, ΔS. With the exception of the first change in magnetization at the Kwak field, as field is increased from zero, the other anomalies appear as discontinuities. The discontinuous jumps persist to temperatures of >1/4 K. The temperature dependence of the fields at which the jumps occur, which serves as a phase diagram for the system is shown in Fig. 4.

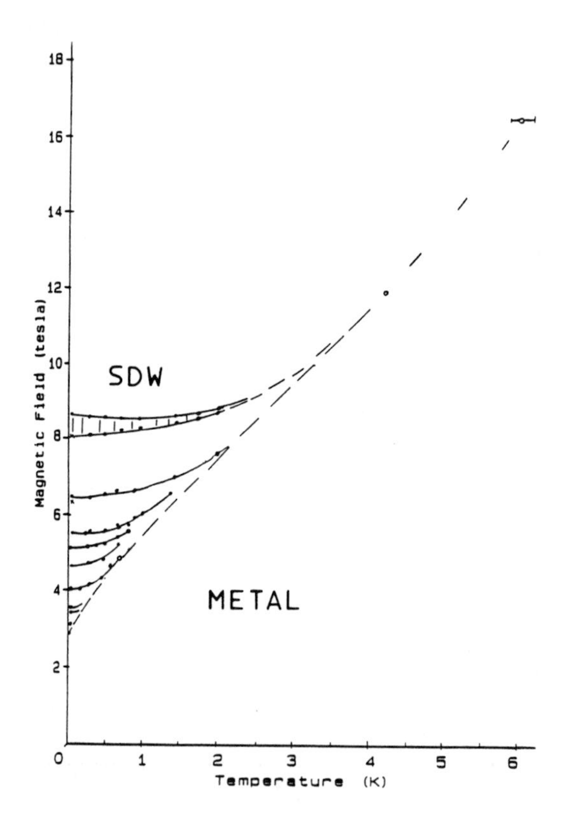

Figure 4. Phase diagram for $(TMTSF)_2ClO_4$. Low field phase is metallic, high field phase is SDW semimetal.

III. A Theoretical Model

After the Hall resistance measurements suggested a series of SDW transitions, a theory was needed to explain how a magnetic field could induce a phase transition into an SDW state. The experiments had shown that the effect was orbital rather than spin since the transitions only occur when the field is perpendicular to the highly conducting a-b planes. Orbitally induced transitions are rare in nature.

The physics behind the transition should have been clear from Ted's earlier paper discussed above [1]. One dimensional metallic systems are intrinsically unstable against Fermi surface instabilities such as SDW's or CDW's (charge density wave) which lead to a semiconducting state. The instability results from the the fact that in one dimension all of the degenerate states at the Fermi energy are coupled by the same wavevector, $q=2k_f$. Thus an infinitessemal distortion at this wavevector drastically alters the nature of the wavefunctions. We have seen that a two dimensional open orbit metal in a magnetic field becomes one dimensionalized in the presence of a perpendicular magnetic field. This is the basic cause of the field induced transition, but it took several years for the explanation to be discovered [9,10]. The argument also shows that two dimensional open orbits are unstable at zero temperature since any magnetic field will induce a density wave transition.

Once it was known that the system becomes unstable the question was how to treat the new phase. Conventionally for SDW or CDW transitions one introduces a lattice or spin distortion of wavevector $q\sim 2k_f$, calculates the resulting gap near the Fermi energy and adjusts the amplitude of the gap to minimize the total energy, electronic + lattice (the elastic or spin wave energy involved in creating the distortion). For free electrons in one dimension the calculation of the spectrum requires solving

$$-(\hbar^2/2m)(\partial^2/\partial x^2)Y + \Delta\cos(qx)Y = E(k_x)Y \tag{3}$$

A gap opens at the q/2 and states with lower k have their energy lessened. Integrating up to E_f the total electronic lowering is $(N(E_f)\Delta)(\Delta \ln(\Delta/E_f))/2$, where N(E) is the density of states. The elastic energy cost is propotional to Δ^2/K, where K involves the elastic constants and the electron-distortion coupling. The electronic term always wins due to the logarithmic term and minimizing with respect to Δ one arives at the mean field form for Δ and T_c.

$$T_c \sim \Delta \sim T_f \exp(-1/N(E_f)K) \tag{4}$$

In the present case the problem is not so trivial since the equation for the electronic spectrum already contains a periodic term. By analogy to the simple treatment above we add a distortion term to arrive at:

$$-(\hbar^2/2m)(\partial^2/\partial x^2)Y + 2t_b \cos(2\pi x/\lambda)Y + \Delta\cos(qx) = E(k_x)Y \tag{5}$$

where $\lambda = hc/eHb$. We now have a Shrodinger equation with two potentials, generally incommensurate, one very large ($2t_b \sim 300K$) but slowly varying ($\lambda \sim 1000$ Angstroms at 10 Tesla) and the other small ($\Delta \sim T_c \sim 1K$) but rapidly varying ($1/q \sim 1/2k_f \sim 5$ Angstroms).

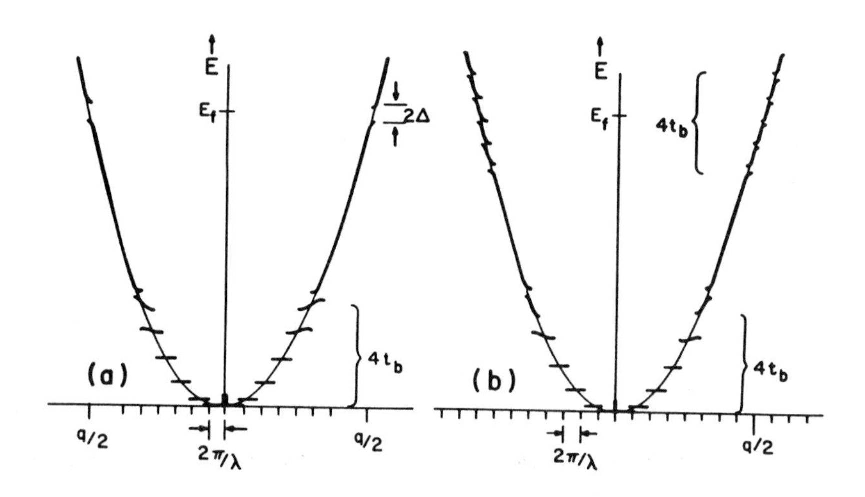

Figure 5. Dispersion relations for a) the gaps from the two potentials superposed and b) the actual gaps when both potentials are present.

Naively one might expect that the potentials have little to do with one another since the first produces gaps up to $4t_b$ and gives a continuous spectrum at E_f while the second produces a small (Δ) single gap at $\sim E_f$ if taken alone as shown in Fig. 5a. However, the actual solution is far more interesting than the superposition of the gaps in different energy regions [11]. In fact the combination of the potentials yields a series of gaps about $E_0(q/2)=\hbar^2(q/2)^2/2m$ over an energy interval $4t_b$ wide as illustrated in Fig. 5b. As one leaves this region the gaps decay exponentially, much as the gaps in the solutions to Mathieu's equation decay exponentially for energies larger than the amplitude of the cosine potential.

The long wavelength period associated with the magnetic field and the coefficient t_b, shows up in this diagram as the spacing between the gaps ($2\pi/\lambda$ along k, or $\hbar\omega_c'$ along E) and therefore the number of gaps is $4t_b/\hbar\omega c'$. Rigorously there are, of course, gaps at every k value $n\pi+m\pi/\lambda$ (m and n integers), but most of these gaps are exponentially small giving rise to the continuous bands in this region (as well as outside). The only gaps which survive are those shown. Thus only the period $2\pi/\lambda$ is observable in this region. The amplitude of the gaps varies with magnetic field as is illustrated in fig. 6a with the form

$$E_m = \Delta|J_m(4t_b/\hbar\omega_c')| \qquad (6)$$

which is approximately equal to

$$E_m = (\Delta/(4t_b/\hbar\omega_c')^{1/2})|\cos(4t_b/\hbar\omega_c'-m\pi/4-\pi/4)| \qquad (7)$$

This latter form holds for the case when we choose any periodic potentials with periods 2π and $2\pi/\lambda$. The Bessel function form, J_m, is explicitly the solution when the periodic potentials are cosines. Note that the variation of the gap amplitude with field is such that adjacent gaps are out of phase, so that one is maximum when the adjacent gaps are zero.

If we were to choose $q=2k_f$, the gap at the Fermi energy would oscillate as the magnetic field is changed and the system would have a series of transitions oscillating between SDW and normal metal states. This was the scenario introduced in the pioneering paper of Gor'kov and Lebed [9], who only found and considered the effects of the single gap at the Fermi energy. However, the experiments do not show any reentrance of a normal state between the different SDW transitions. The resolution of the problem is clear when one looks at the complete spectrum shown above. While the gap at the Fermi energy is decaying to zero the gaps above and below are reaching their maximum value. The system can therefore gain more electronic energy by switching the wavevector of the distortion from q to $q-2\pi/\lambda$, or $q+2\pi/\lambda$, which brings the larger gaps in coincidence with the Fermi energy [11].

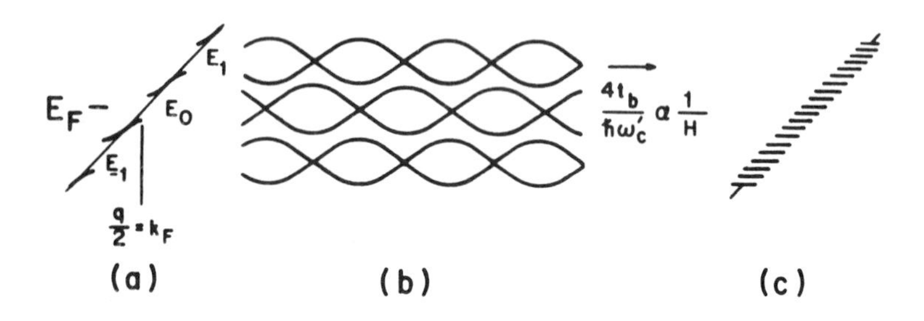

Figure 6. a) Details of the gap structure near E_f. b) Oscillations of the gaps as H changes., c) The limit where $\lambda \to \infty$.

The changing wavevector for the distortion would happen well before the gap reached zero amplitude, and since the order parameter for the different wavevector distortions is finite when the wavevector jumps, this would correspond to a series of first order transitons as is seen experimentally. Thus the presence of many gaps around the Fermi level and the series of phase transitons with changing wavevector are inherent in the problem of the field induced transition for open orbit two dimensional metals.

The idea of changing the distortion wavevector in order to arrive at a series of different SDW phases was introduced previously by a number of authors who also included more details of the bandstructure and allowed the distortion wavevector to vary in q_y as well as q_x [12,13]. However, the presence of a multitude of gaps was not evident in the earlier works [14].

The spectrum for incommensurate potentials often yields surprising results. In the present case one of the biggest surprise occurs when one allows the long wavelength, λ, to approach infinity. At first glance one would expect this limit to correspond to a 'flat' potential which has no effect except to uniformly shift the energies. However, what actually occurs is that the bandwidths associated with all of the bands in the region $4t_b$ wide about E(q) become exponentially small while the number of such bands increases algebraically. Thus the entire region is filled with gaps, the bands having zero width, Fig. 6b. The spectrum appears as if it has one large gap of width $4t_b$ around the energy given by E(q). This resembles what one might expect if the amplitude of the slowly varying potential were interchanged with that of the rapidly varying potential.

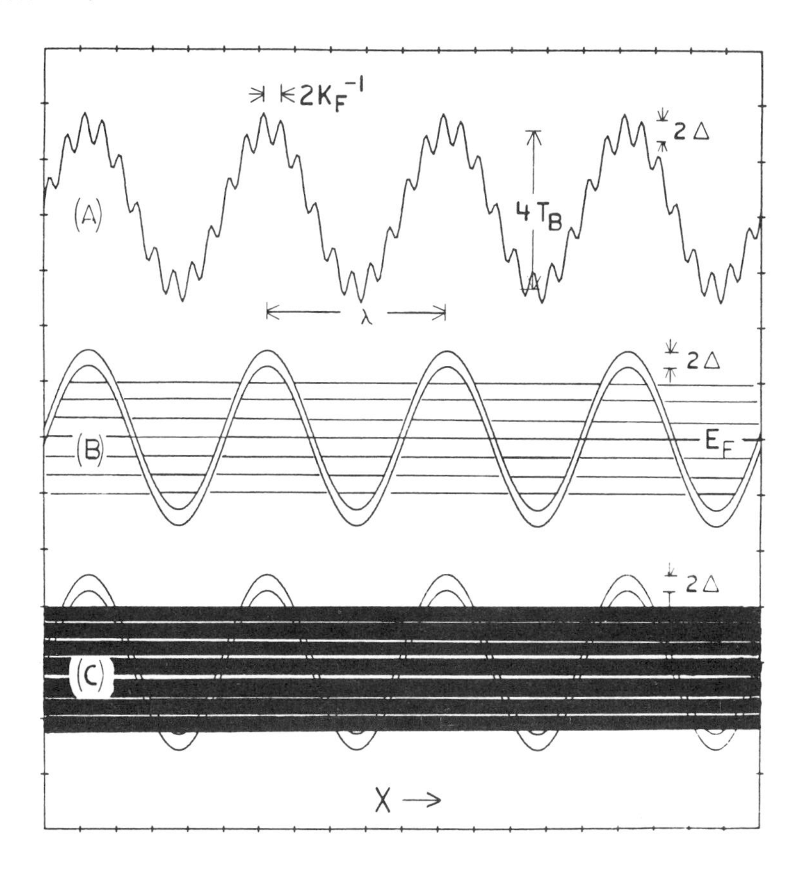

Figure 7. A) The potential in real space. B) The effect of treating the fast potential and quasiclassical quantization. C) Allowing tunneling, the discrete levels form bands.

Qualitatively one can understand this progression by looking at a quasi-classical approximation, although the actual calculation requires a detailed mathematical treatment. The combined potential from the two terms is shown in Fig. 7a. In quasi-classics one treats the rapidly varying potential first. This leads to a gap 2Δ at the Fermi energy which is then modified by the slowly varying potential as is shown in Fig. 7b, in treatment very similar to the way one handles band bending in a semiconductor. We can then quantize the electron and hole states in the potential wells in the top and bottom of the valence and conduction bands and arrive at discrete energy levels. The tunneling between these discrete levels broadens them into bands and qualitatively gives the picture shown in Fig. 6. As λ is increased the spacing between the discrete levels in Fig. 7c decreases algebraically but the tunneling through the barriers decreases exponentially so that the bandwidths shrink exponentially.

IV. Saturns Rings

We sent these calculations off for publication and an anonymous referee suggested that the problem we were treating had been looked at before by Avron and Simon[15]. The cited paper indeed treated the same differential equation (equation 5) but with x replaced by t (time). The physical problem they were addressing was an explanation of the anomalously large gaps in the rings of Saturn, particularly the size of the Cassini division. It had previously been suggested that the gaps in ring systems are due to resonance orbits. If a particle in a ring is resonant with another larger body orbiting the same planet (or star) (that is the particle has an orbital frequency which is rationally related to that of the larger object) then the orbit that the particle sits in is unstable and a gap will develop [16]. In the case of the Cassini division the frequency of an orbit with a radius inside the division has twice the frequency of the small moon Mimas. However, estimates of the size of the gap that should result from this resonance yield a gap only ~3 kilometers, whereas the actual width is 4500 kilometers.

Astronomers had suggested that the large width might be caused by the presence of sheparding moons at the edges of the Cassini division, but high quality voyageur photographs failed to detect the presence of these moons. Avron and Simon suggested an alternative explanation. Along with the perturbation due to Mimas they suggested that there was a perturbation due to the rotation of Saturn around the sun. Mimas is light (3.76×10^{19} kilograms), and has a period of .942 earth days. The Sun is mammoth (1.989×10^{30} kilograms) and the period of the Saturn year is 10,759.22 earth days. Even taking into account the differences in the distances for the gravitation interaction (distance Mimas - Cassini division ~ 7×10^4 kilometers, distance Saturn - Sun ~ 1.4×10^9 kilometers, ellipticity of Saturns orbit ~ 10 %), Mimas gives a small rapidly varying perturbation while that of the sun is large and slowly varying.

Avron and Simon used perturbation theory to treat the combined potentials in the limit which corresponds to $4t_b/\hbar\omega_c' \ll 1$. They found some small additional gaps but nothing spectacular. In the appropriate limit $4t_b/\hbar\omega_c' \gg 1$ they did not solve the problem but conjectured that the combined effects of the incommensurate potentials would produce a spectrum like that found by Hofstadter [4], with zero measure for the allowed bands and a large region filled by gaps. Since they not solved the problem they did not have a way of estimating the magnetude of the enhancemant of the resonance gap.

The solution to this incommensurate problem as we have outlined above, is not precisely what Avron and Simon had envisioned but it is close enough that their conjecture rings true. In the limit when $4t_b/\hbar\omega_c' \gg 1$ we have seen that the spectrum yields extremely narrow bands and a region filled only with gaps. Can these results explain the Cassini division? If the mapping of the orbital stability problem to the Shrodinger form is appropriate (as has not yet been demonstrated) then using the numbers for the masses and distances described above, one of us (P.M.C.) finds that the resonance gap from Mimas alone maybe amplified by a factor of ~1000 when the perturbation from the sun is included. Taking the previous estimate of the size of the resonace gap at 3-20 kilometers [15,16], this certainly puts the calculated gap in the ballpark of the observed gap in the Cassini division. However, it should be noted that more recent work [16] does not even mention [15] and there are other theories relying on the effects of spiral density waves.

V. Remaining Problems

From the depths of space we now return to the low temperature high magnetic field world of the organic conductors. The basic instability which leads to the field induced transitions and even the series of first order transitions are reasonably well understood. What is left to understand? Alot! All of the models suggest that after the 'last' transiti n, presumably at ~ 8 Tesla at low temperature the materials should be insulating. They are not. They remain semimetallic to much higher fields. Recent experiments suggest that there maybe a new, possibly insulating transition in the region of 25-30 Tesla [17]. One of the most puzzling newer discoveries, was the observation initially by Ribault [18], of an additional transition and a large reversal of the sign of the Hall coefficient right in the middle of the cascade of phase transitions, at 6 -7 Tesla, but only when the samples were very slowly cooled through the anion ordering temperature of 24K. The slow cooling allows the anions more time to order and results in a more perfect crystal with a longer mean free path. This is reminiscent of the calculation by Thouless [19] of the effects of Landau subbands on the Hall coefficient. Thouless showed that in traversing the Hofstadter diagram, which is filled with gaps of all sizes, the Hall coefficient could make 'wild' swings in magnetude and sign. The biggest effects come from the smallest gaps, and one might expect to be able to see them only with extremely clean samples. We do not expect that the Hofstadter problem is directly applicable, since we do not have a square lattice [20]. It merely serves as a guide as to what might happen when one does a full quantum treatment of the present problem. To date only quasi - classical approximations have been attempted with the exception of the model presented here, which neglects the Brillouin zone in the x or a direction. There are in fact three competing periods, the magnetic length, $1/2k_f$ and the lattice parameters. The possiblity of splitting the Landau levels by a combination of other periods is quite intriguing but awaits further experimental and theoretical work.

VI. Conclusion

The paper by Holstein, Chaikin and Azbel, although a minor contribution by Ted's standards started us on the road to understanding not only a very interesting problem in low temperature condensed matter physics, but also some of the complexities of the spectrum of incommensurate potentials and possibly a solution to a centuries old problem, that of the Cassini division in Saturn's rings.

VII. Acknowledgements

We acknowledge many useful interactions with Ralph Chamberlin, Mike Naughton, Xiao Yan, Jim Kwak and Jim Brooks. Research support is from NSF DMR85-14825 and LRSM DMR82-16718.

1. P. M. Chaikin, T. Holstein, and M. Ya. Azbel, Phil. Mag. B48, 457, (1983).
2. J. F. Kwak, J. E. Schirber, R. L. Greene, and E. M. Engler, Phys. Rev. Lett. 46, 1296 (1981) and Mol. Cryst. Liq. Cryst 79, 121 (1981)
3. M. Ya Azbel, Zh. Eksp. Teor. Fiz. **46**, 939 (1964) [Sov. Phys. JETP **19**, 634 (1964)],
4. D. R. Hofstadter, Phys.Rev. B14, 2239 (1976).
5. P. M. Chaikin, Mu-Yong Choi, J. F. Kwak, J. S. Brooks, K. P. Martin, M. J. Naughton, E. M. Engler, and R. L. Greene, Phys. Rev. Lett. 51, 2333 (1983).
6. M. Ribault, D. Jerome, J. Tuchendler, C. Weyl, and K. Bechgaard, J. Phys. Lett. 44, L-953 (1983).
7. M. J. Naughton J. S. Brooks, L. Y. Chiang, R. V. Chamberlin and P. M. Chaikin, Phys. Rev. Lett. 55, 969 (1985).
8. F. Pesty, P. Garoche and K. Bechgaard, Phys. Rev. Lett. 55, 2495 (1985).
9. L. P. Gor'kov and A. G. Lebed, J. Physique Lett. 45, L-440 (1984).
10. P. M. Chaikin, Phys. Rev. B31, 4770 (1985).
11. M. Ya Azbel, Per Bak and P. M. Chaikin, to be published in Phys. Rev. A
12. G. Montambaux, M. Heritier, P. Lederer, J. Phys. Lett. 45, L-533 (1984), M. Herritier and P. Lederer, Phys. Rev. Lett. 55, 2078 (1985).
13. K. Yamaji, J. Phys. Soc. Japan 54, 1034 (1985), K. Yamaji, Synthetic Metals, 13, 29 (1986).
14. A. Virosztek, L. Chen and K. Maki, to be published in Phys. Rev., K. Maki, to be published in Phys Rev.
15. J. E. Avron and B. Simon, Phys. Rev. Lett. 46, 1166 (1981).
16. F. Franklin, M. Lecar and W. Wiesel, In Plnetary Rings, ed. by R. Greenberg and A. Brahic, (The University of Arizona Press, Tucson 1984) p.562
17. R. V. Chamberlin, M. J. Naughton, J. S. Brooks, X. Yan and P. M. Chaikin, to be published.
18. M. Ribault, Mol. Cryst. Liq. Cryst. 119, 91 (1985)
19. D. J. Thouless, M. Kohmoto, M. P. Nightingale, and M. den Nijs, Phys. Rev. Lett. 49, 405 (1982).
20. J. F. Kwak, J. E. Schirber, P. M. Chaikin, J. M. Williams, H.-H. Wang and L. Y. Chaing, Phys. Rev. Lett. 56, 972 (1986).

Today's Small Polaron*

David Emin

1 INTRODUCTION

In 1959 Ted HOLSTEIN published his initial two papers about polarons [1,2]. The first of these companion papers concerns the formation of a polaron. The second addresses the motion of a small polaron. These papers have provided a clear conceptual base upon which to build. Nonetheless, within the quarter-of-a-century since their appearance, understanding of small-polaron formation and motion has changed considerably. The present paper describes a current view of how small polarons form and move. In addition, an interesting group of materials, the boron carbides, are taken to illustrate a recently observed novel feature of transport in polaronic systems.

First, the theory of small-polaron formation is discussed. It is emphasized that we now know that self-trapping is dichotomous. That is, except for truly one-dimensional systems, a charge carrier in a solid with a short-range electron-lattice interaction either self-traps to form a small polaron or does not self-trap at all. In addition, I emphasize that the imposition of modest disorder can trigger self-trapping. An analytically soluble example illustrates how the imposition of relatively small site-to-site (diagonal) disorder induces the abrupt collapse of a large-radius state to a severely localized small-polaronic state.

Second, the semiclassical theory of small-polaron hopping is discussed. Early treatments viewed small-polaron hopping as proceeding via a succession of independent jumps in which a charge carrier moves between adjacent sites [2]. However, often the successive jumps of a carrier are "correlated" [3-5]. Then, hopping occurs in flurries during which there are many hops. These flurries are followed by dormant periods during which the carrier remains static. Here, the physical origin of this "correlated" hopping motion is described. The transient motion of a carrier injected into a narrow-band molecular solid is then considered [6]. Upon insertion, an injected carrier can make nonactivated hops between adjacent sites before the atoms surrounding the carrier equilibrate about positions consistent with the presence of the carrier, i.e., form a true small polaron. The mobility associated with this transient motion is ~ 1 cm^2/V-sec. It is only after the atomic vibrational motion equilibrates about an essentially static carrier that the very low, thermally activated mobility generally associated with small-polaron hopping motion occurs. The equilibration time can be sufficiently long that time-of-flight mobility measurements can manifest relatively high activationless mobilities while steady-state measurements reveal the low thermally activated mobility characteristic of a small polaron.

Finally, the electronic transport properties of the boron carbides are briefly discussed. The boron carbides are very distinctive systems in which the charge carriers form "small" singlet bipolarons. I describe the features of the boron carbides which enable small-bipolaron formation to occur.

2 SMALL-POLARON FORMATION

Self-trapping is a nonlinear process. To illustrate the basic self-trapping phenomenon, consider a static carrier introduced in a deformable solid. For simplicity, the electronic energy of a charge at a site depends only on the strain at that location: a short-range electron-lattice interaction. Here, the presence of a static charge carrier exerts a force on the atoms it encounters. This force acts to displace these atoms so as to lower the energy of the electron. As a result, the electron becomes increasingly localized. With increased localization, the cancellation of the forces exerted on an atom from different regions of the electron's density is reduced. Thus, the net (electron-lattice) force a static charge exerts on a given atom in its immediate vicinity increases as the electronic wavefunction is compressed. This increased force, in turn, increases the severity of the localization. Here we see the feedback of self-trapping: increased localization augments the effective electron-lattice interaction which then enhances the localization.

As a result of the nonlinearity associated with this feedback effect, there is a dichotomy between extended (or weakly localized) states and severely localized small-polaronic states.

Mathematically, this dichotomy is seen from the nonlinear wave equation describing the adiabatic groundstate of an electron in a three-dimensional deformable continuum with a short-range electron-lattice interaction [7,8]:

$$[-\hbar^2\nabla^2/2m - (F^2a^3/k)\ |\psi(\underline{r})|^2]\psi(\underline{r}) = E\psi(\underline{r}). \qquad (1)$$

Here F is the force exerted by a severely localized electron on the atoms adjacent to it (the electron-lattice coupling strength), a is the inter-atomic separation, k is the lattice stiffness (Hooke's-law constant), m is the carrier's mass and $\psi(\underline{r})$ and E are the carrier's groundstate eigenfunction and eigenvalue, respectively. The key feature is that the potential energy term mirrors the electronic density, $|\psi(\underline{r})|^2$.

$|\psi(\underline{r})|^2$ is of the order of V^{-1}, where V is the volume of the electronic eigenstate. If the carrier were completely delocalized, $V \to \infty$, the carrier-induced potential well of (1) vanishes. Thus, extended states are free-particle-like. However, if the carrier is severely localized, $V \approx a^3$, the potential energy of the electron is reduced by a substantial amount, $\sim F^2/k$, the self-trapping potential for a small polaron. Thus, we see that the wave equation of the adiabatic groundstate for an electron in a deformable continuum admits two very different solutions.

To show the dichotomy (that the adiabatic groundstate is either severely localized or extended), we can proceed further with a scaling argument [7,8]. First, one calculates the groundstate energy by multiplying (1) by $\psi^*(\underline{r})$ and integrating over $\underline{r}$:

$$E = (\hbar^2/2m) \int d\underline{r}\ |\nabla\psi(\underline{r})|^2 - (F^2a^3/k) \int d\underline{r}\ |\psi(\underline{r})|^4. \qquad (2)$$

We note that the groundstate wavefunction for an electron in an isotropic continuum must be spherically symmetric. Thus, the first term of Eq.(2) scales with spatial extent, R, as R^{-2}, while the second term varies as R^{-3}. This means that the first term, the kinetic energy term, dominates when the radius is made large, $R \to \infty$. However, as the radius shrinks to zero, $R \to 0$, the potential energy term dominates. As a result, a plot of the energy versus the wavefunction's spatial extent is a simple <u>peaked</u> function as illustrated in curve a of Fig 1. The curves of Fig. 1 have no meaning when the spatial extent of the wavefunction shrinks below a, a value comparable to the interatomic separation. Hence we only consider the curves for $R \geqq a$. The minimum energy solution thus either occurs at $R = \infty$, corresponding to a delocalized electron with negligible electron-induced atomic displacements, or occurs at $R = a$, corresponding to a severely localized solution for which there are substantial electron-induced atomic displacements. Owing to the translational degeneracy of a solid, the first situation gives rise to the usual band of Bloch states, while the second gives rise to a small-polaron band [2,7]. These two types of states represent the essential dichotomy of self-trapping.

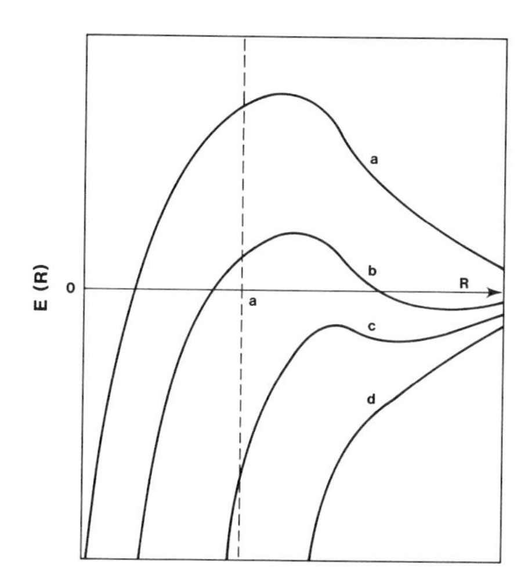

Fig. 1 The groundstate energy of the system comprising an electron in a deformable medium attracted by a Coulombic center is plotted against the radius of the electronic state, R. Curves a through d illustrate the effect of increasing the strength of the attractive Coulombic force [8].

In the presence of an additional localizing potential, as that due to a Coulombic center associated with a donor state, the possible groundstates of the coupled electron-lattice system remain dichotomous [8]. Namely, the presence of a Coulombic center provides an additional potential energy term in (2) which is proportional to $1/R$. The groundstate is either of large radius, $R \gg a$, or severely localized, $R = a$. As the strength of the localizing potential of the Coulombic center is increased (curves a, b and c of Fig. 1), the energy of the small-radius state at $R = a$ is lowered relative to the large radius minimum at $R \gg a$. Ultimately, in curve d, only the severely localized minimum exists. The severely localized state is small-polaronic. That is, the well-localized carrier induces substantial displacements of the atoms about it.

It is clear that driving forces for localization, such as the previously described Coulombic potential, foster the formation of a small-polaronic state. Since small-polaron formation is dichotomous and disorder fosters localization, one may expect disorder to abruptly trigger the formation of small-polaronic states.

It is often quite difficult to readily control and characterize disorder. However, the magnetic disorder of a ferromagnet well below the Curie temperature is controlled with temperature and describable in terms of spin waves. Hence, it is interesting to ask whether increasing the temperature of a ferromagnetic semiconductor is associated with any localization phenomena. In fact, it is known that with rising temperature, large-radius donor states in the ferromagnetic semiconductor EuO abruptly collapse to severely localized states at about 2/3 the Curie temperature [9]. Here I present a model in which this thermally induced donor-state collapse is triggered by the increased generation of spin waves [10].

The magnetic interactions of the donor electron with the local spins of the magnetic semiconductor provide the feature which distinguishes a donor in a magnetic solid from the usual donor state, described previously. In particular, an electron of spin $\underline{s}$ senses the spins of the Eu atoms ($S = 7/2$) through the exchange interaction. The exchange interaction of an electron of spin $\underline{s}$ on site $\underline{g}$ with the localized spin of that site is $-2I\underline{s}\cdot\underline{S}_{\underline{g}}$, where I is the exchange interaction energy. In EuO the electronic motion is associated with an intersite electronic transfer energy which greatly exceeds $|I|$. In this situation, the electron finds it energetically favorable to maintain its spin direction as it moves between sites [11]. Designating this direction as the z-direction, we rewrite the interaction energy as $-2Is^zS^z_{\underline{g}}$. With inequality of the orientation of the local spins, the electron experiences a disorder potential as it moves between Eu-sites.

We compute the expectation value of the electron-magnon interaction with a hydrogenic electronic wavefunction of radius R. Then, we express the deviations of the local spins from saturation in terms of the presence of spin waves. One finds that the contribution of the electron-magnon interaction to the system's energy is quadratic in the magnon creation and annhilation operators. As a result, the donor electron induces an R-dependent shift of the magnon frequencies [10]. The R-dependence arises because a donor electron of radius R only interacts appreciably with magnons with wavelengths in excess of ~4R.

In equilibrium, the system comprising the donor electron interacting with the phonons and magnons will be that which minimizes the free energy. In particular, to address the thermally induced abrupt collapse of a large-radius donor state to a severely localized state, we focus our attention on the R-dependent terms of the free energy. In the absence of the electron-magnon interaction, we have the situation described in Fig. 1. Namely, the kinetic-energy, electron-lattice and Coulombic contributions are respectively proportional to R^{-2}, $-R^{-3}$ and $-R^{-1}$. For sufficiently weak (screened) attraction of the donor electron to the donor center and a sufficiently weak electron-lattice interaction, illustrated in curve b of Fig. 1, the stable donor state is of large radius, R >> a. With the addition of the electron-magnon interaction, there is a shift of the magnon frequencies induced by the presence of the donor electron. This frequency shift contributes a term to the free-energy which increasingly favors localization as the temperature, T, is raised: $\propto -TR^{-2}$. As illustrated in Fig. 2, with increasing temperature the free-energy minimum associated with the severely localized state is lowered relative to that of the large-radius donor state. At the temperature of the collapse of the donor state, the free-energy minimum of the severely localized state falls below that of the large-radius donor.

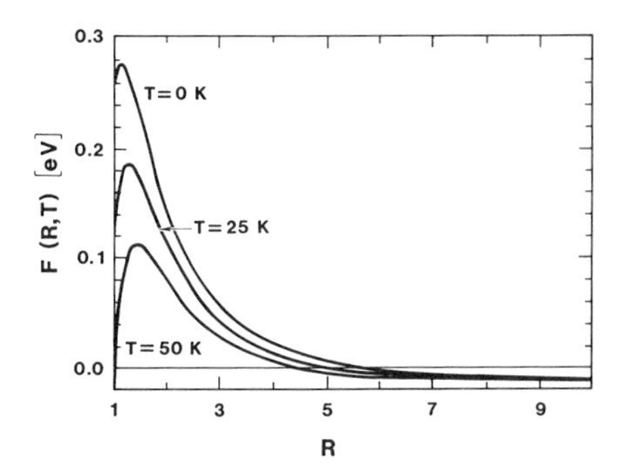

Fig. 2 The R-dependent portion of the free energy of a donor electron interacting with atomic displacements and spin waves is plotted against a dimensionless donor-state radius, R, at three temperatures [10]. There is a shallow minimum at R ≈ 10.

It is important to realize that the thermally induced abrupt donor-state collapse is the result of the synergistic effect of two driving forces for localization: the polaron effect and the spin disorder (characterized by the electron-lattice and electron-magnon interactions). The presence of the short-range electron-lattice interaction produces the dichotomous situation required for the abrupt collapse. Indeed, without the electron-lattice interaction, our treatment yields only a small continuous shrinking of the donor-state wavefunction with increasing temperature. Furthermore, the presence of the electron-magnon interaction produces a temperature-dependent driving force for localization, without which there is no thermally induced transition. In other words, in our treatment the thermally induced abrupt donor-state collapse is the result of the combined localizing effects of the electron-lattice and electron-magnon interactions. Individual consideration of these effects produces qualitatively different results.

Finally, it is emphasized that the thermally induced donor-state collapse differs from Anderson localization. For the donor-state collapse, the imposition of modest disorder triggers the formation of a small-polaronic state. In fact, the electronic disorder energy is the smallest energy in the problem. Numerical estimates of the electronic disorder energy at the transition temperature yield values of less than 0.1 eV [10]. Indeed, the largest energies in the problem are the electronic bandwidth and the small-polaron binding energy. As is common [12], for the donor-state collapse, we have near-cancellation of the respective delocalizing and localizing effects associated with these two energies. It is the magnetically induced disorder which tips the near-balance to produce sudden and extreme localization. More generally, we should not be surprized when the relatively modest electronic disorder characterizing many noncrystalline solids induces the dramatic change from high-mobility nonpolaronic motion to low-mobility small-polaronic behavior [13].

3 SMALL-POLARON HOPPING MOTION

The standard picture of small-polaron hopping motion (and phonon-assisted hopping motion, generally) [2] views successive hops as independent events. Namely, with the absorption and emission of phonons an equilibrated carrier is transported to another site at which equilibrium of the carrier with the phonons is reestablished. Holstein provided us with a clear semiclassical picture of high temperature ($T \geq$ the phonon temperature) small-polaron hopping: the occurrence probability approach [2]. Here, we first use this semiclassical picture to explore whether successive hops of a small polaron are independent of one another. Then, a picture of "correlated" small-polaron hopping motion is described.

In Holstein's occurrence probability approach one envisions an electronic carrier initially equilibrated on a site. In Fig. 3a, the site is taken, as in Holstein's molecular crystal model [1,2], to be a molecule characterized by a single deformation parameter. In our illustration, with occupation by a charge carrier, the equilibrium value of this deformation parameter is increased. As a result, the electron energy level associated with a carrier on a molecule at its equilibrium configuration is lowered with respect to that of an undeformed molecule. In the semiclassical approach, motion to another site (molecule) requires the two local electronic energy levels to be momentarily degenerate with one another. This degeneracy can be achieved with appropriate deformations of the molecules involved in the hop. When the lattice deformations are such that the two local electronic energy levels involved in the hop achieve a momentary energy degeneracy, we have a "coincidence event," Fig. 3b. With this coincidence, the carrier can move to another site. The standard hopping scheme is completed when the carrier equilibrates with the molecular vibrations at its newly occupied site, Fig. 3c.

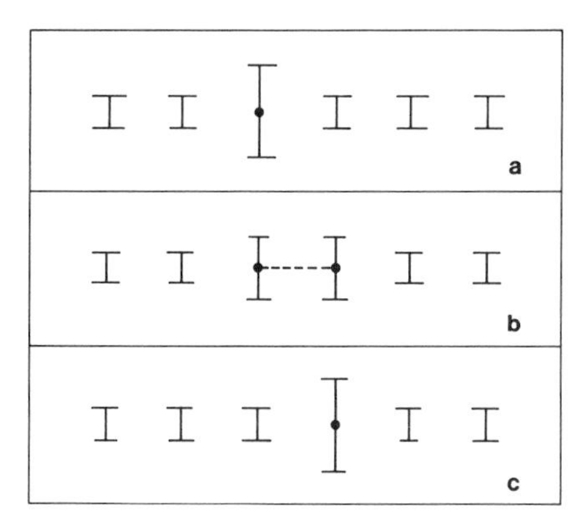

Fig. 3 A schematic representation of the small-polaron jump process in the high-temperature semiclassical regime. Nonessential "thermal" displacements and vibrational dispersion effects are suppressed for clarity.

The jump rate is the product of (1) the rate characterizing the formation of the coincidence event and (2) the probability that the carrier will be able to avail itself of the opportunity to move and thereby negotiate a hop, P. The minimum energy which must be supplied to deform the "initially" and "finally" occupied molecules so as to form a coincidence is the activation energy for hopping, E_A. If the electronic transfer integral linking inital and final molecules is sufficiently large [2], the charge transfer probability for a given coincidence event is essentially unity ($P \approx 1$). Then the hopping is referred to as "adiabatic." If the electronic transfer integral is small enough (not a common occurrence), the electron cannot always follow the lattice motion and avail itself of the opportunity to hop ($P \ll 1$). The hopping is then said to be "nonadiabatic."

With the formation of a coincidence configuration, the involved molecules are vibrationally excited. Reequilibration involves dispersing this energy, $\sim E_A$, to the solid as a whole. At long times, $\Delta\omega\, t \gg 1$, the residual energy remaining in the molecules involved in the coincidences is $\sim E_A(\Delta\omega\, t)^{-d/2}$, where $\Delta\omega$ is the vibrational dispersion (the width of the band of optical frequencies associated with the molecular vibrations) and d is the dimensionality of the system [3.4]. In other words, the coincidence is a "hot spot" and vibrational relaxation corresponds to its cooling to the ambient temperature. Thus, the relaxation time τ is determined by the relation $E_A(\Delta\omega\, \tau)^{-d/2} = k_B T$, where k_B is the Boltzmann constant.

Immediately after a hop, before vibrational relaxation is complete, the electron finds itself in an athermal environment in which the probability of hopping is enhanced over that for a carrier on a relaxed molecule. Indeed, immediately after a carrier negotiates a hop it is typically confronted with a substantial probability of returning to its previously occupied site [3-5]. That is, an "immediate" return hop is very likely. In addition, because the region surrounding the carrier is "hot," the carrier also has an augmented probability of making a hop to yet another site. Thus, successive small-polaron hops are expected to be correlated unless the probability of a carrier responding to a coincidence event by making a hop is so small ($P \lll 1$) that relaxation will generally take place before a secondary hop occurs. Hence, "correlated" small-polaron hopping is expected to be a common occurrence.

When small-polaron hopping is "correlated," we can view the jumps as occurring in flurries. Namely, there are periods of frequent hopping followed by dormant periods. During the periods of frequent hopping there is a large probability of a hop being followed by a return hop. Thus, one can view a carrier as shuttling back-and-forth between a pair of sites. Actual diffusion occurs when the carrier alters its shuttling motion from one pair to another pair which shares one common site (e.g. from 1-2 to 2-3). In thermal equilibrium, these periods of rapid motion are compensated by periods in which the carrier remains stationary. Thus, despite quite a different physical picture, one obtains an average jump rate equal to that calculated assuming the carrier to equilibrate after each hop [2].

The "correlated" hopping of a small-polaron can, however, be observed in a transient experiment where the electron-lattice system is not in equilibrium. For example, consider an electron injected into a narrow-band molecular crystal [6]. Here we envision a carrier with a deBroglie wavelength comparable to an intersite separation. That is, with its small spatial extent, we can regard the carrier as confined to a single molecule at any given time.

Let us consider what happens when an injected carrier is placed on a molecule. Since the molecular distortions associated with small-polaron formation are generally very much larger than those of thermal displacements, a carrier will light on a relatively undeformed molecule. With the arrival of the carrier, the molecule will experience a force driving it to achieve its small-polaronic atomic displacement pattern. In essence, the placing of the carrier on a molecule provides an impulse to the molecule's atoms. Since the atomic displacements associated with small-polaron formation are generally very much larger than those of vibrations, the arrival of the carrier induces large athermal vibrations about the carrier-induced equilibrium configuration. In particular, the carrier's arrival induces molecular vibrations with an energy comparable to the small-polaron binding energy, E_b, where $E_b >> k_BT$.

As illustrated in Fig. 4, with a single deformational mode per molecule, the vibrational energy imparted to the occupied molecule is typically sufficient for it to achieve coincidences with a neighboring molecule at later times. In the Holstein Molecular Crystal Model, one deformational mode per site, these times are approximate multiples of a vibrational period. These "extra" coincidences lead to an enhanced hopping rate for the carrier before vibrational relaxation occurs. Calculating the rate for a hop at time t after the carrier is placed on a molecule with no more than thermal distortions, one finds a time-dependent jump rate [6]. Explicitly, one has

$$R(t) = \nu P \exp\{-E_A[1 - G(t)]^2/k_BT\}, \tag{3}$$

where ν (= $\omega/2\pi$) is the vibrational frequency of the molecule's deformational mode, E_A is the hopping activation energy in a relaxed lattice, $E_A \sim E_b/2$ [2], and G(t) is a function which describes the relaxation of the lattice following a hop [5]. Explicitly, the relaxation function is

$$G(t) = A(\Delta\omega\ t) \cos[\omega t + \phi(\Delta\omega\ t)]. \qquad (4)$$

It contains two factors: 1) an amplitude factor which depends on the intermolecular vibrational coupling (the dispersion of the phonon frequencies) and 2) an oscillatory factor which varies with a frequency $\sim\omega$ modified by a phase factor which varies relatively slowly (with a time characteristic of intermolecular energy transfer $\sim 1/\Delta\omega$). The first factor falls from unity to zero with time as the lattice relaxes. The second factor reflects the oscillatory dynamics of the coincidence situation. Namely, a coincidence at $t = 0$ tends to lead to another at $2\pi/\omega$, while leaving little probability of a coincidence at $t = \pi/\omega$. Thus, the time-dependent activation energy of (3), $E_A[1 - G(t)]^2$, vanishes at $t = 0$, $[G(0) = 1]$. It oscillates with time with an amplitude which decreases with time until the activation energy becomes the time-independent valued, E_A, when G(t) falls to zero.

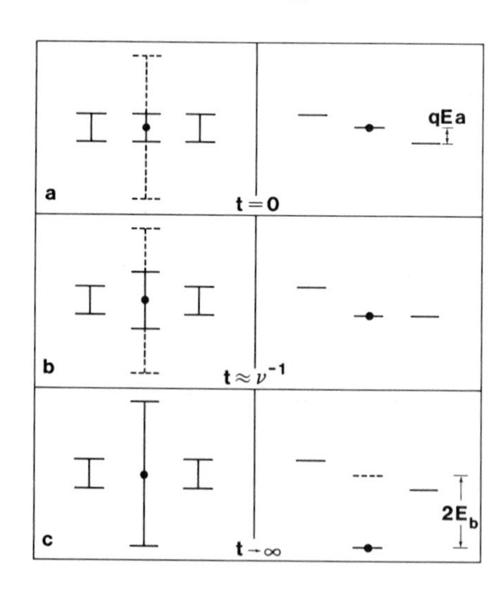

Fig. 4 A schematic representation of the configurational and energetic relaxations which follow the placing of a carrier on the central molecule of a row of three equivalent molecules at $t = 0$. The left-hand column depicts the displacments of a configurational coordinate of each of the three molecules from their carrier-free equilibrium positions. The right-hand column depicts the electronic energy levels associated with each of the three molecules in the presence of an electric field of strength E. With the carrier's charge being q and the intemolecular separation being a, the potential energy difference between adjacent molecules is qEa. For simplicity, thermal fluctuations of the molecular distortions and the concomitant energy levels are ignored. Subfigure a illustrates the situation at $t = 0$. When the carrier arrives at the central molecule, the molecule is at its carrier-free equilibrium configuration. However, upon arrival of the carrier, the equilibrium value of the configurational coordinate of the occupied molecule is displaced to the new value shown by the dashed line. Subfigure b depicts the situation

after about one vibrational period, $t \approx 1/\nu$. The configurational coordinate of the occupied molecule has begun to relax toward its equilibrium value. Concomitantly, its electronic energy falls. Here the electronic energy level at $t \approx 1/\nu$ is shown to be in coincidence (degenerate) with that of a neighboring molecule. Hence the carrier has an opportunity to make an activationless transfer to that molecule. Subfigure c shows the fully relaxed (equilibrium) situation. The occupied molecule is displaced to the equilbrium configuration consistent with the presence of the carrier while the carrier's energy is reduced by $2E_b$, a value twice the small-polaron binding energy, E_b. The dashed level in c shows the unrelaxed electronic energy level.

Relaxation occurs when the minimal activation energy for hopping rises to a value comparable to k_BT. This relaxation requires a time, $t' = (6/\Delta\omega)(k_BT/E_A)^{1/4}$. The probability for a carrier to hop to another site in a system with six neighbors before the lattice relaxes is $1 - \exp(-6\nu Pt')$. For adiabatic hopping ($P = 1$) the term in the exponential is generally much greater than unity. Thus, there is an extremely high probability of a carrier hopping from one slightly distorted site to another before lattice relaxation can occur. In other words, on average a carrier will make many such hops before it remains on a site long enough for full vibrational relaxation to occur. For the Holstein model, (a narrow-band molecular solid with only one deformational mode per molecule) times as long as 1 sec are possible [6]. Prior to vibrational relaxation the carrier hops in a nearly activationless manner with a mobility $\approx |q|\nu a^2/k_BT \approx 1$ cm^2/V-sec, where q is the carrier's charge and a is the lattice constant.

With such long relaxation times in ideal systems, one expects the defects and impurities of a real system to play an important role in the equilibration of a carrier. That is, with a typical estimate of trap densities, it is almost a certainty that a carrier will encounter a trap before it relaxes at an intrinsic site. Since trapped carriers remain at a site longer than untrapped carriers, they have an enhanced probability of achieving vibrational relaxation before making another hop. That is, defects and impurities can serve as equilibration centers. Then, the transport picture is that the carrier will move rapidly until it relaxes in a defect or impurity-related state. Its subsequent motion is the low, thermally activated hopping motion characteristic of an equilibrated small polaron.

Thus, an equilibrated small polaron is generally associated with a low, thermally activated mobility. However, its transient mobility can be much higher and not thermally activated. These effects can be significant in time-of-flight and transient photoconductivity experiments.

4 APPLICATIONS OF SMALL-POLARON THEORY

In the early days of small-polaron theory there was little experimental evidence of small-polaron formation. However, small polarons have now been found in many systems. These include transition-metal oxides, molecular crystals, glasses, ionic insulators, rare-gas solids, liquids and dense vapors. In addition, essential features of small-polaron theory have been applied to the problem of light interstitial diffusion in solids [14,15], electronically stimulated desorption from surfaces [16], and trapping and release of carriers from traps in semiconductors [17]. Here, I shall describe several distinctive features of polaronic effects in the boron carbides.

Boron-rich borides, generally, and the boron carbides, in particular, are characterized by distinctive bonding and structures [18-21]. For example, the building blocks of the boron carbides are twelve-membered boron-rich-clusters of atoms. In each cluster, there are thirty lines connecting each atom with its neighbors. If these lines were interpreted as two-center bonds, sixty (2 X 30) electrons would be required to fill the internal bonding requirements of a boron icosahedron. However, it has been demonstrated that only 26 electrons are required to fulfill the internal bonding orbitals of the icosahedron [18]. Only 26 electrons are required because the internal bonding of these icosahedra, so called "electron deficient bonding," is of the three-center type [19]. Namely, rather than having the charge density of bonding electrons centered along the lines joining adjacent atoms, the charge is located at the center of the equilateral triangles which make up the faces of the icosahedron. On average, there is a charge density of 1.3 electrons on each of the 20 faces of an icosahedron.

With each of the atoms of a boron icosahedron bonded to an atom external to the icosahedron via a standard two-center bond, as in a borane molecule ($B_{12}H_{12}$), there are only 24 electrons from the second shell of the boron atoms available for an icosahedron's internal bonding: 12 atoms (3 electrons/atom) - 12 electrons for external bonding. Thus, each such boron icosahedron is desirous of adding two electrons to fill its bonding orbitals. In fact, it is energetically favorable for the borane molecule to be charged doubly negative. With the addition of the two electrons the icosahedron contracts due to the increased strength of its internal bonding [22]. This is just the polaronic effect.

As shown in Fig. 5, in boron carbides, boron-rich icosahedra are linked via direct bonds and through intermediate three-atom linkages. The two additional electrons needed to fill the internal bonding orbitals of boron icosahedra are provided in two ways. First, carbon atoms may substitute for boron atoms within an icosahedron. Thus, a $B_{11}C$ icosahedron requires only one additional electron to fill the needs of its internal bonding orbitals. Second, three-atom chains, such as a C-B-C chain, have one unbonded electron available for donation to a boron icosahedron.

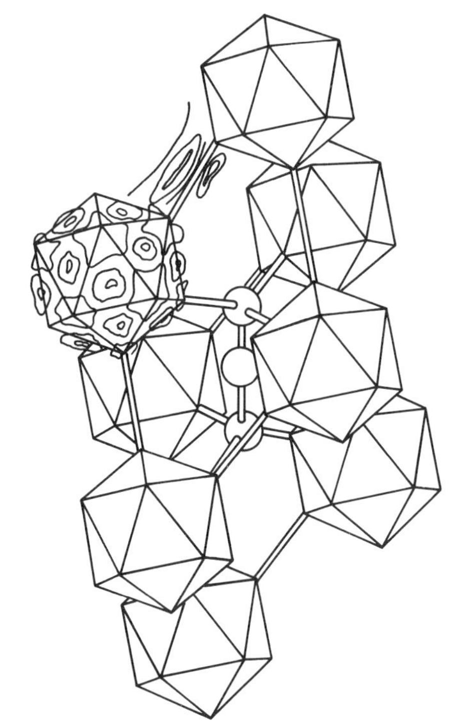

Fig. 5 The structure of boron carbides.

Single-phase boron carbides exist with carbon concentrations from less than 9% to up to 20%. With a range of carbon concentrations, the bonding orbitals are generally only partially filled. The wide composition range is made possible by there being a distribution of $B_{11}C$ and B_{12} icosahedra and a distribution of C-B-C and C-B-B chains [20]. Thus, while maintaining its crystallographic order, boron carbides are compositionally disordered. It may be this feature which causes the electronic charge transport in these materials to be of low mobility.

The electronic transport of boron carbides between 10% and 20% carbon has been studied extensively [23]. Here, the density of charge carriers, holes in the internal bonding orbitals of boron icosahedra, is primarily determined by carbon concentration. The carrier mobilities are found to be no more than weakly dependent on carbon concentration. Mobilities are estimated from the electrical conductivity and the estimated carrier density. In addition, the Hall mobility is measured directly. As illustrated in Fig. 6, the electronic transport is just what is expected of adiabatic small-polaron hopping [24]. Namely, both the high-temperature electrical conductivities and Hall mobilities are thermally activated. The activation energy of the mobility which enters into the conductivity is about 0.15 eV.

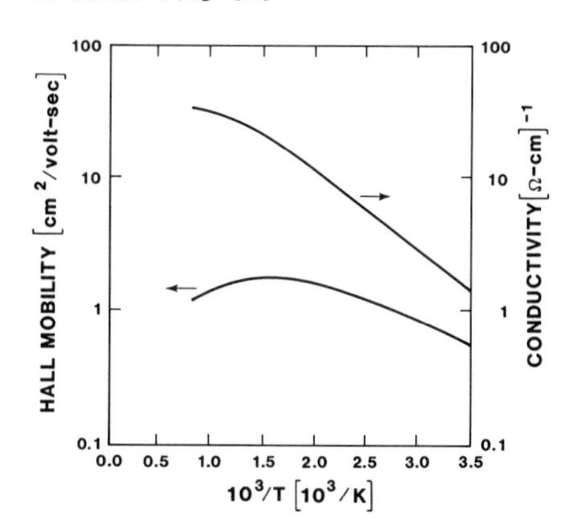

Fig. 6 The electric conductivity and the Hall mobility of a boron carbide with 20% carbon, B_4C, is plotted against reciprocal temperature.

Measurements of the density of paramagnetic spins generally yields temperature-independent values which are orders of magnitude lower than the estimated carrier density [25]. Alternatively put, if the measured spin density was a measure of the carrier density, the mobilities would 1) be orders of magnitude larger than that which is measured, and 2) their temperature dependences should be those characteristic of high mobility solids rather than being the observed hopping type. Thus, it is concluded that the charge carriers are not the origin of the spins. In fact, the spins are thought to be caused by carbon atoms which occupy central positions in the three-atoms intericosahedral chains.[26]. This implies that the charge carriers are spinless. They are small <u>bipolarons</u> [20,25,27].

The origin of small-bipolaron formation in the boron carbides is readily understood in terms of the chemistry of boron-rich icosahedra. Namely, recall that a boron icosahedron is stabilized by the addition of two electrons [18,19,22] which fill the icosahedron's internal bonding orbitals. Here the additional bonding energy gained by adding the two electrons overcomes their Coulomb repulsion. One reason that this inequality occurs is that the electronic charge of the two electrons is distributed over the "surface" of the twelve-atom icosahedral unit. Thus, the magnitude of the Coulomb repulsion is greatly reduced from that associated with the occupation of a single atomic site.

For small-bipolaron formation to occur in a solid, the lowering of the energy of the system associated with two charges being added to a site (icosahedron) must be greater than twice the energy lowering associated with single-site occupancy. In the boron carbides the energetically favorable site for the occupation of two bonding electrons is a $B_{11}C^+$ icosahedron [20,22,27]. In a $B_{11}C^+$ icosahedron, the positive charge provided by the carbon nucleus helps offset the Coulomb repulsion of the two added electrons. Explicitly, here small-bipolaron formation presumes that it is energetically favorable for the system to form a single $B_{11}C^-$ icosahedron rather than two $B_{11}C$ icosahedra:

$$2B_{11}C \rightarrow B_{11}C^+ + B_{11}C^- . \tag{5}$$

The energy lowering associated with the addition of an electron to a $B_{11}C^+$ icosahedron to form a $B_{11}C$ icosahedron is the small-polaron binding energy, E_b. The addition of two electrons to a $B_{11}C^+$ icosahedron leads (with a linear electron-lattice interaction) to a polaronic lowering of the energy of the site of $4E_b$. The factor 4 arises because the addition of a second electron doubles the magnitude of a carrier-induced atomic deformation pattern experienced by each of the two added electrons. With the inclusion of a Coulombic energy U for the interaction between the two electrons added to the $B_{11}C^+$ icosahedron, the energetic relation required for small-bipolaron formation becomes:

$$-2E_b > -4E_b + U. \tag{6}$$

Small bipolaronic hopping occurs when pairs of electrons hop between $B_{11}C$ icosahedra:

$$B_{11}C^- + B_{11}C^+ \rightarrow B_{11}C^+ + B_{11}C^-. \quad (7)$$

Aside from small-bipolaron formation the boron carbides display a number of novel polaronic properties described and explained elsewhere [20,23,28]. These effects include a huge disorder-induced enhancement of the high-temperature Seebeck coefficient and a pressure-induced reduction of the hopping conductivity.

5 SUMMARY

We now know that self-trapping is dichotomous in multidimensional systems. A charge carrier will self-trap to form a small polaron or it will remain essentially free. This sharp distinction reflects the non-linearity of self-trapping phenomena. Furthermore, we have seen how various driving forces for localization act synergistically to produce small-polaron formation. Namely, modest disorder can act with the electron-lattice interaction to trigger small-polaron formation.

We have come to see that most small-polaron hopping is adiabatic. In these instances successive hops of a small polaron are often correlated. Hopping transport occurs in flurries. This type of sporadic motion prevails when the time for the lattice to relax following a hop is very long compared with the mean hopping time. Despite the presence of correlations between successive hops, the steady-state hopping of a small-polaron remains characterized by low, thermally activated mobilities predicted by the early studies of small-polaronic transport. Correlation effects do, however, manifest themselves in transient situations when an injected carrier has a substantial probability of moving between sites before the lattice can relax and thermalize in a manner appropriate to its occupation of a particular site. In these instances, an injected carrier can be associated with a relatively high mobilitiy (≥ 1 cm^2/V-sec) which can persist for a long time in ideal circumstances. Thus, a photoconductivity experiment may not manifest the characteristic small-polaron transport even though a steady-state transport experiment will.

Small-polaronic effects have become an important topic in solid state physics. Self-trapping has been found in many systems. Solids in which small bipolarons are formed have even been found. Boron carbides provide an example of distinctive systems in which we can understand why the charge carriers form small bipolarons.

Since Ted Holstein's initial work on polaron formation and small-polaron motion, there has been considerable growth of our understanding of polaronic phenomena. Ted has played a key role in many of these efforts. Here only a few topics have been addressed. Indeed, such important topics as the Hall effect and the motion of one-dimensional polarons have not been discussed. Ted's influence cannot be measured by his publications alone. Ted has also contributed greatly through the very many insights he has passed on through discussions. His presence will be deeply missed. He will not be forgotten.

6 REFERENCES

1. T. Holstein: Ann. Phys. (N.Y.) 8, 325 (1959)
2. T. Holstein: Ann. Phys. (N.Y.) 8, 343 (1959)
3. D. Emin: Phys. Rev. Lett. 25, 1751 (1970)
4. D. Emin: Phys. Rev. B 3, 1321 (1971)
5. D. Emin: Phys. Rev. B 4, 3639 (1971)
6. D. Emin, A. M. Kriman: Phys. Rev. (in press)
7. D. Emin: Adv. Phys. 22, 57 (1973)
8. D. Emin, T. Holstein: Phys. Rev. Lett. 36, 323 (1976)
9. J. B. Torrence, M. W. Shafer, F. R. McGuire, Phys. Rev. Lett. 29, 1168 (1972)
10. D. Emin, M. Hillery, N. H. Liu: Phys. Rev. B 33, 2933 (1986)
11. P. W. Anderson, H. Hasegawa: Phys. Rev. 100, 675 (1955).
12. Y. Toyozawa: Prog. Theor. Phys. 26, 29 (1961)
13. D. Emin: Comm. Sol. State. Phys. 11, 35, 59 (1983)
14. C. P. Flynn, A. M. Stoneham: Phys. Rev. B 10, 3966 (1970)
15. D. Emin, M. I. Baskes, W. D. Wilson: Phys. Rev. Lett. 42, 791 (1979)
16. D. R. Jennison, D. Emin: Phys. Rev. Lett. 51, 1390 (1983)
17. D. V. Lang, R. A. Logan, M. Jaros: Phys. Rev. B 19, 1015 (1979)
18. H. C. Longuet-Higgins, M. de V. Roberts: Proc. Roy. Soc. London, A230, 110 (1955)
19. W. N. Lipscomb: J. Less-Common Metals 82, 1, (1981)
20. D. Emin: In Boron Rich Solids, ed. by D. Emin, C. L. Beckel, T. L. Aselage, I. A. Howard, C. Wood (American Institute of Physics, New York, 1986)
21. D. Emin: Physics Today (in press).
22. I. A. Howard, C. L. Beckel, D. Emin: In Boron Rich Solids, ed. by D. Emin, C. L. Beckel, T. L. Aselage, I. A. Howard, C. Wood (American Institute of Physics, New York, 1986)

23. C. Wood, D. Emin: Phys. Rev. B 29, 4582 (1984)
24. D. Emin, T. Holstein: Ann. Phys. (N.Y.) 53, 439 (1969)
25. L. J. Azevedo, E. L. Venturini, D. Emin, C. Wood: Phys. Rev. B 32, 7970 (1985)
26. D. Emin, E. L. Venturini: Bull. Am. Phys. Soc. 31, 300 (1986)
27. D. Emin, G. A. Samara, L. J. Azevedo, E. L. Venturini, H. H. Madden, G. C. Nelson, J. A. Shelnutt, B. Morosin, M. Moss: J. Less-Common Metals 117, 415 (1986)
28. G. A. Samara, D. Emin, C. Wood: Phys. Rev. B 32, 2315 (1985)

*Supported by U.S. DOE under Contr. DC-AC04-76DP00789.

Electron-Phonon Coupling

T. H. Geballe

I'm here today to represent the silent majority of solid state physicists who were not fortunate enough to experience the thrill of giving a seminar to a Holstein-containing audience. The challenge of his tough persistent questions apparently kept people awake and at the same time raised the quality of the research. I enjoyed talks with Ted at Physical Society Meetings and had planned to try to interest him in thinking about possible models of enhanced electron-phonon interactions in superconductors at the next opportunity.

Ted's published papers have long time constants. Consider, for instance the Phys. Rev. Lett., (1954) in which he accounted for the then-unexpected-spectral-dependence Biondi found in optical absorption in metals. He showed by second order perturbation that emission and absorbtion of phonons resulting from the interaction of the photon field, the Bloch electrons, and the lattice, would account for the observed spectral behavior found in metals and the temperature dependence. It took until the 1970's before these Holstein-processes were observed directly. Richards' group at Berkeley and Sievers' group at Cornell did so using thin superconducting films. Analysis of those far infrared measure

-ments gave the Eliashberg function, $\alpha^2(\omega)F(\omega)$, the product of the phonon density of states $F(\omega)$ weighted by the electron-phonon coupling, $\alpha(\omega)$. The Eliashberg function is of paramount importance in strongly coupled superconductivity of which I speak today.

The Eliashberg function is now more easily obtained by tunneling spectroscopy than by optical means. It is found by analyzing the differential conductance of S/I/N (superconducting-insulating- normal) tunnel junctions. Methods developed by McMillan and Rowell for doing this proved to be quantitative ways of unfolding the Eliashburg equations for Pb and other non-transition metals soon after Ivar Giaever discovered superconducting tunneling. Lead oxidizes at its surface and forms a good oxide barrier which is essential for obtaining quantitative results. Advances in vacuum deposition techniques now make it possible to deposit amorphous Si, Al, (and other elements) as overcoats which can be oxidized to form pinhole-free artificial barriers. The application of tunneling spectroscopy to superconductors which do not form good native oxide barriers is now common place using these artificial barriers. In particular, with the allowance for a proximity-effect due to a monolayer or so of degraded superconductivity at the interface, refractory elements and compounds such as Nb_3Ge, can be dealt with using the method of Wolf and Arnold.

The strong-coupling Eliashberg theory has been used as a stimulus for experimental approaches to reaching higher T_c's, unfortunately without much visible success. The highest generally accepted T_c is an onset temperature of just over 23 K in sputtered Nb_3Ge films, found by Gavaler, and by Testardi well over a decade ago. As the concentration of Ge is increased towards the stoichiometric amount, the T_c rises and the films become metastable. The onset temperature depends upon the type of measurement used in following the superconducting transition. If the measured quantity is resistance the onset temperature is a measure of the first detectable continuous path of superconductivity and may in fact represent less than 1% of the films. In cases where there is an inhomogeneous compositional distribution which is filamentary in nature it can be considerably less. In many cases there is a much broader and lower range of T_c's corresponding to the compositional distribution throughout the film. The Eliashberg functions for a series of NbGe samples is shown in Fig. 1.[1] A small leakage current detected near zero bias in the tunnel current of the highest T_c Nb_3Ge film is not large enough to prevent the analysis from being carried out self consistently. The leakage is taken as evidence of the presence of the second phase which is just starting to nucleate and is not yet in sufficient concentration to be detectable by more conventional means. Evidently in the highest-T_c, highest-gap junctions the metastability limit for the synthesis of the A15 phase on the Ge-rich boundary is slightly exceeded and a trace of material has transformed to the more Ge-rich stable sigma phase which is not superconducting. The transformed material has a finite density of states to tunnel into at 0 bias, hence the leakage current.

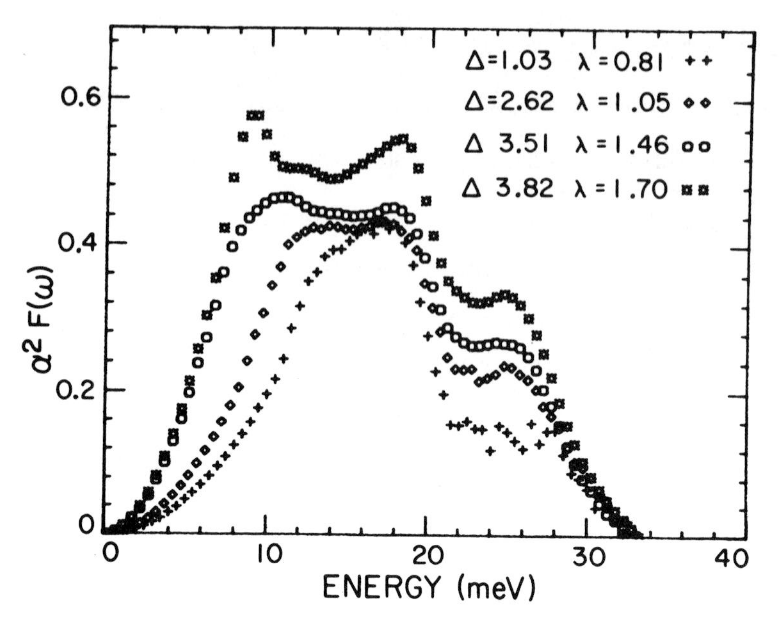

Fig. 1--Taken from Reference 1. The 4 curves are for NbGe films with the A15 structure (in descending order) containing 24.7, 24.3, 21.8 and 16.7 percent Ge with T_c onsets of, respectively, 21.2K, 20.1K, 16.8K and 7.0K.

How does the Ge-rich sigma phase nucleate? One possibility is by a simple displacement reaction. The so-called Hyde-rotation is an eigenvector which takes the A-15 structure into the sigma phase.[2] It is then possible that the phonon-softening evident in Fig. 1 provides a path for the nucleation which then allows the disproportionation reaction into a germanium-rich (σ phase and germanium-poor (A15) phase to proceed.

Integration of the Eliashberg function shown in Fig. 1 gives λ = 1.83, $2\Delta/k_BT_c$ = 4.17 for the NbGe sample at the limit of metastability. Here λ, the attractive electron-phonon interaction, is the ω^{-1} moment of the Eliashberg function and Δ is the energy gap at low temperatures. The employment of processing strategies for increasing the range of metastability i.e., increasing the Ge concentration in the A15 phase, such as rapid quenching, surface epitaxial stabilization, high pressure synthesis, and chemical or interstitial gas doping, which different researchers throughout the world have been employed, has had little effect.

The increase of $2\Delta/kT_C$ above the mean field BCS value = 3.5 is also a measure of the degree of strong-phonon coupling. In Fig. 2 we can see for 3 different systems that the ratio rises as a function of concentration to approximately the same value before the limit of metastability is reached.

From data like that shown Figs. 1 and 2 there is empirical evidence that the raising of T_C above ~ 18 K is accompanied by mode softening (i.e. an increase in the low frequency portion of the Eliashberg function) and an increase in λ. For the high T_C vanadium-based A15 compounds; $T_C \leq$ 17 K, on the other hand the experimental results are much different. No evidence of mode-softening or enhanced strong coupling is evident. The superconductivity in V_3Si is due to a high electronic density of states and high Debye temperature. Most transition metal superconductors including V_3Si behave in accordance with the Varma-Dynes tight binding theory in which the electron density of states is the parameter which most directly affects T_C

The low values of $2\Delta/kT_C$ (< 3.5) seen in Fig. 2 were at one time taken as a signal that the analysis might be flawed and therefore the empirical results discussed above were suspect. We now know that the

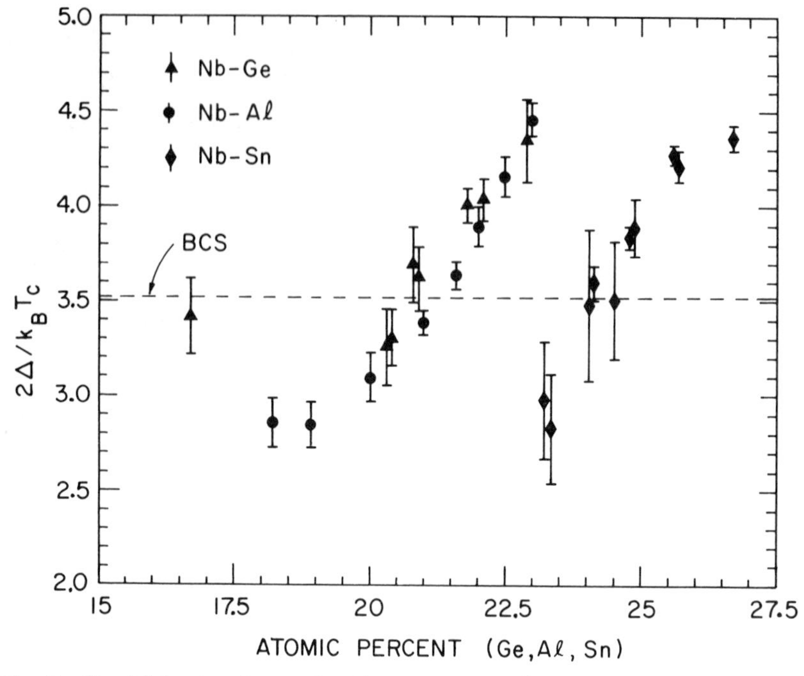

Fig. 2--The highest values of ratio of the gap (measured at low temperatures) to the onset T_C are for the highest obtainable concentrations of Sn, Al, or Ge in the A15 phase.

low values are the result of measuring T_C as an onset temperature and Δ as an averaged gap.[3] The discrepancy between the onset and average quantitites is exacerbated by concentration gradients which form in the off-stoichiometric samples presumeably by surface diffusion during film growth[4]. The heat capacity measures the transition of the entire film averaged over length scales larger than the superconducting coherence length (50Å). Manifestations of concentration gradients can be seen in Fig. 3 for an off-stoichiometric sample. The transition for the stoichiometric sample is, like that measured electrically, quite sharp. In contrast, the heat capacity transitions for off-stoichiometric samples are as broad as 5 K demonstrating the presence of macroscopic concentration gradients, even though the electrically measured transitions in the same samples are less than one tenth as wide. Evidently paths of connected superconducting filaments are formed only a few tenths of a degree below the temperature where the first onset region is detected. It is assumed of course that the distribution of T_c is due to concentration gradients. Values of $2\Delta/kT_c < 3.5$ are indicative of inhomogeneous films.

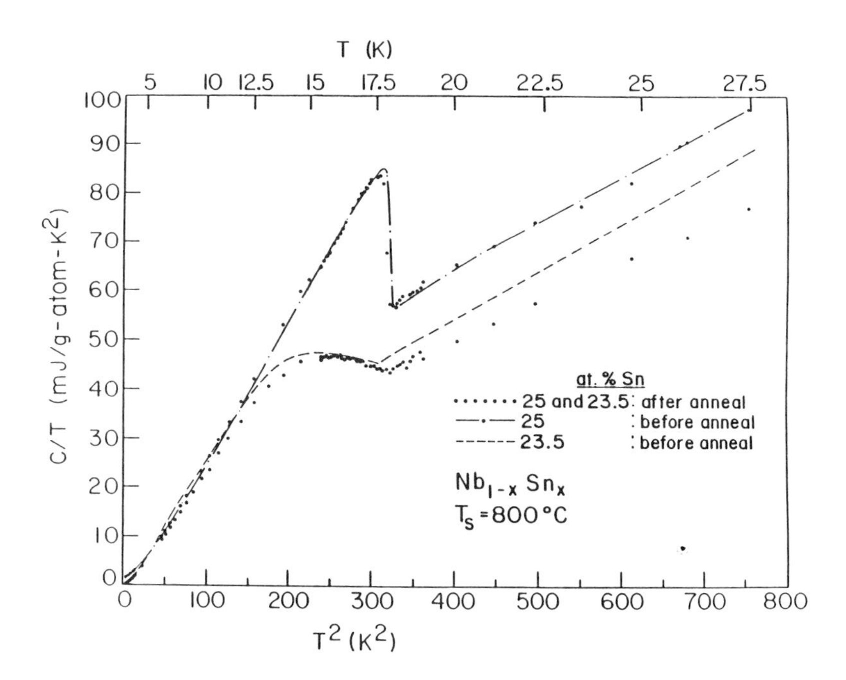

Fig. 3--Heat capacity transitions for a stoichiometric (25% Sn) and a non-stoichiometric (23.5% Sn) sample both before and after annealing for 24 hours at 800 degrees C. The transitions measured electrically (by inductive or resistive) in comparable samples would be sharp in both cases

A recent report by Ogushi and Osono[5] leads me once again to review the possibility of synthesizing systems with enhanced superconductivity. Advances in vacuum deposition make it possible to deposit films with unusual interfaces which was not possible just a few years ago. Enhanced superconductivity was suggested by Ginsburg and modeled for a specific system by Allender Bray and Bardeen (ABB) some years ago. Anderson and Inkson have used arguments to demonstrate that the ABB model will not enhance superconductivity. As discussed by Cohen and Louis, the arguments do not preclude enhanced superconductivity if umklap processes or local field corrections become important. Unfortunately the small average-gap of PbTe, the semiconductor used in the ABB model because it leads to a large electronic polarizability, also implies that the local field correction is small.

Starting with Little's one-dimensional excitonic model a large variety of other mechanisms have been suggested for obtaining large enhancements of superconductivity. The models develop electron-electron attractive interactions based upon the exchange of bosons. Plasmons, excitons, d-mons (acoustic plasmons) and electron- hole pairs have all been considered. None of these have been realized experimentally in any convincing way. Sham has shown that the estimates of T_C are all much too high because the validity of Migdals' theorem has been assumed. I would like to suggest a still different mechanism, which might have applied to the Ogushi results had they been valid, but anyway may be realizeable with modern thin film techniques. The model requires the sharp interface as envisoned by Ginsburg and ABB, but now between a high T_C superconductor and a polarizeable oxide, specifically an A15 superconductor with a transition metal oxide such as V_2O_5 or Nb_2O_5. The latter form small polarons when a test charge is inserted in the lattice. The electron-phonon coupling constant for V_2O_5, for example, is estimated as ~ 17.[6] It is conceivable that Nb_2O_5 can grow on the surface of the $Nb_{3+x}Ge$ crystallites by outward diffusion of the excess Nb driven by the free energy of formation of the oxide. In fact, Ihara has observed by XPS measurements in some Nb_3Ge samples an enriched Nb (oxide) surface with an underlying depleted Nb layer.

At the interface the quasi-particles in the superconductor would tunnel into the oxide. Some might be trapped and form small polarons. On the other hand if not trapped, they would remain itinerant quasi-particles in the superconductor; their fluctuating fields would induce positively charged responses in the oxide, retarded in time, which could enhance Cooper pair formation.

It should be remarked here that polaronic superconductivity itself is not a new idea. Zinanon[7] building on the Holstein picture of small polarons as quasi-particles moving in narrow bands has shown that the unusual superconducting behavior observed in $SrTiO_3$ can be explained by a model based on disperson of the lowest transverse optical (ferroelectric) mode. The model explains the unusual maximum in $T_c \sim 1K$ found in doped $SrTiO_3$ for carrier concentrations two orders or more below typical metallic carrier densities. Other explanations have also been offered for this intriguing maximum at such a low carrier found many years ago by Schoolery and coworkers. Cohen suggested intervalley phonons. Appel invoked soft modes associated with the 110 K distortion of the lattice. The interface model we have discussed could also explain the results if it were found the reduced or doped $SrTiO_3$ is inhomogeneous on a length scale which varies with concentration, and is, of course small. There is no evidence that such is the case at present but I am not sure it has been looked for with sufficient care.

Finally, I would like to consider results of Sung Park at Stanford which show that it is now feasible to use tunneling spectroscopy to study very thin films and interface superconductivity. Park has made good tunnel junction with films of Nb down to 9 Å i.e. ~ 3 atoms thick. These films are prepared from the vapor phase by epitaxial deposition on single crystal sapphire at room temperature. T_c depends on the thickness -dependent mean-free path exactly as in bulk Nb for thicknesses greater than 50 Å. Below 50 Å the ratio of $2\Delta/kT_c$ increases surprisingly, even to a larger value than for the A15 compounds, as shown in Fig. 4. Although the analysis is preliminary, it is clear that the electron-phonon coupling becomes stronger. By itself this should make T_c increase. In fact, T_c decreases showing that an additional effect becomes dominant. For example, as the films become more two dimensional, reduced screening can cause the (repulsive) Coulomb pseudopotential to increase, although the resistance itself is so low that weak localization effects are not important.

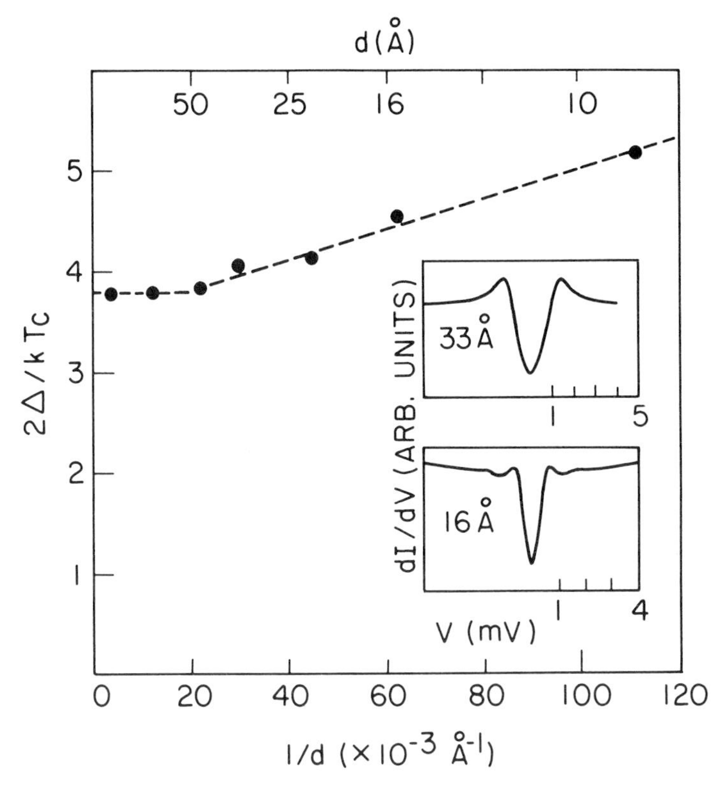

Fig. 4--$2\Delta/kT_c$ as a function of inverse film thickness. The insets show the conductance data for 33Å and 16Å films, *Nb*.

A number of clues remain to be investigated before the idea of enhanced surface or interface superconductivity is either proven or should be dismissed. We should be able to gain a better understanding of the enhanced superconductivity in granular aluminum. Here the cause of the rise in T_c from ~ 1 to ~ 3K as the particle size of Al embedded in Al_2O_3 decreases is not clear even after much study. In another system, single crystal Nb, twinned by at deformation in liquid He, is found to have its onset T_c increase by as much as 2 K[8] suggesting interface phonons may play a significant role. The report of the superconductivity of a monolayer of Ag on [100] surface of Ge is clearly suggestive. It should be possible to synthesize many new interfaces with intriguing properties in the near future. I trust that there are clear thinkers - in the Holstein tradition - around to keep on puncturing ballooons and hopefully to enhance the of enhancement.

REFERENCES

1. K. E. Kihlstrom, D. Mael and T. H. Geballe, Phys. Rev. B 29, 250 (1984); K. E. Kihlstrom, Ph. D. dissertation, Stanford University (1985).
2. Kwo, Geballe, Hammond, AIME Metallurgical paper
3. D. A. Rudman, F. Hellman, R. H. Hammond and M. R. Beasley, J. Appl. Phys. 55, 3544 (1983).
4. F. Hellman, Ph. D. dissertation, Stanford University (1985).
5. T. Ogushi and Y. Osono, Applied Physics Letters 48, 1167 (1986). I would like to thank Professor Ogushi for informing me (private communication) that the unusual occurrence of superconductivity reported to be above 23K upon further investigation, seems to have been due to an error in the measurement of the sample temperature.
6. M. Henry, C. Sanchez, C. R.'kha and J. Livage, J. Phys. C.: Solid State Physics 18 , 6589 (1985).
7. Z. Zinamon, Philosophical Magazine, V. 21, No. 170, pages 347 - 356, February 1970.
8. Brobrov and Zorin, JETP 40, 8 (1984).

The Random Field Problem; Facts and Fiction

V. Jaccarino

Site random fields have profound effects on the ground state properties and critical behavior of magnetically ordered systems. For the random field Ising model (RFIM), it has been proven theoretically that the lower critical dimension $d_\ell = 2$. A variety of experimental studies by the author and his colleagues on RFIM systems have demonstrated the phase transition to be destroyed at $d = 2$ and a sharp, continuous phase transition at $T_c(H)$ to exist at $d = 3$ with new critical exponents corresponding to an effective dimensionality $\bar{d} \simeq 2$; suggestive of a dimensionality reduction $\bar{d} \simeq d-1$. Furthermore, extreme critical slowing down has been observed as $T \rightarrow T_c(H)$. This explains why field cooling through $T_c(H)$ always traps the $d = 3$ system into a nonequilibrium domain state on what are laboratory time scales. The predicted crossover scaling has been observed for the field dependence of all thermodynamic functions and "phase boundaries" in the distinctly different $d = 2$ and $d = 3$ RFIM systems. Contrary to current theory and the results described above, Birgeneau, Cowley and Shirane- from neutron scattering studies alone - have 1) assumed the low temperature, field cooled configuration is always the ground state of the RFIM, 2) "unambiguously" determined $d_\ell \geq 3$ and 3) discovered a "temporal" and/or discontinuous phase transition occurs at $d = 3$. I suggest the errors in all of their conclusions represent a confluence of the limitations of the use of a single experimental technique, not doing the key experiments with that technique, incorrect interpretations and a concommitant use of poor and/or badly characterized materials.

Apologia

In response to the invitation to present a paper at the Theodore D. Holstein Symposium, I looked for a connection, however tenuous, between Ted's work and the random field problem. None was to be found. Despite his prolific output and diverse interests, he saw fit to eschew this field (perhaps wisely so). Thus what is discussed below has no substantive relation to the body of his research but I do hope it captures the spirit of both his striving for clarification and one particular facet of his character; Ted was forthright and outspoken in his views.

I. Introduction and Theoretical Prologue.

The random field problem had a relatively inauspicious beginning a decade ago in the paper of Imry and Ma.[1] They conjectured on the instability of a ferromagnet against domain formation when subjected to a (site) random field, the configuration average $\langle h_i \rangle$ of which vanished but whose variance did not; *i.e.* $\langle h_i^2 \rangle \neq 0$. Their original arguments are appealingly simple though profoundly important.

Consider a region of a ferromagnet of linear size L containing N spins all pointing "up" upon which is imposed a random field with magnitude $h \equiv \langle h_i^2 \rangle^{1/2}$. Because of statistical fluctuations in h_i there will now be $(N)^{1/2}$ more fields pointing "down" than "up" in this volume $V \sim L^d$, where d is the space dimension. Reversal of all spins in this region will result in a random-field-energy gain $\varepsilon_m \sim hL^{d/2}$ since $N \sim L^d$. But this lowering of the energy is exacted at the cost of raising the interface (or wall) exchange energy ε_w between oppositely pointing spins. The wall-energy increase depends on the spin-dimensionality n of the system. For systems with continuous (e.g. Heisenberg or X-Y) symmetry $n \geq 2$, $\varepsilon_w \sim JL^{d-2}$ but for the Ising ($n = 1$) case $\varepsilon_w \sim JL^{d-1}$ where J is the exchange coupling. Balancing the gain ε_m against the loss ε_w, the wall energy dominates for $n \geq 2$ at $d > 4$, whereas for $n = 1$ it does so at $d > 2$, ensuring that the lower critical dimension d_ℓ of continuous symmetry and Ising systems is $d_\ell = 4$ and $d_\ell = 2$, respectively. By lower critical dimension one means that, at or below d_ℓ, the system is unstable against domain formation however small is h, for arbitrarily large L, and a transition to a long range ordered state (LRO) would not occur, even at $T = 0$ K. The simple Imry-Ma domain argument is not conclusive as to what happens precisely at $d = d_\ell$.

Two facts of life set bounds on further interest in the random field problem. The first is that "real" systems have $d \leq 3$; hence comparison between experiment and theory for $n \geq 2$ systems would be limited to verifying that d is always less than d_ℓ in those instances. The second limitation, which applies to all spin and space dimensionalities, is that we do not know how to generate a site random field in a ferromagnet! Is the whole of the random field then little more than a theorist's conjecture, incapable of experimental verification? Absolutely not!

Without doubt, the most important theoretical paper from the experimental point of view to follow upon Imry-Ma was that of Fishman and Aharony (F-A).[2] F-A showed that the randomly diluted antiferromagnet in a uniform field H, applied colinearly with the direction of spontaneous ordering mapped directly on to the random field Ising model (RFIM) for the ferromagnet. Later, Cardy[3] showed the equivalence to be exact, in the weak field ($h/J \ll 1$) limit, in the sense that both systems belong to the same universality class; an important point in the interpretation of critical phenomena in d = 3 RFIM systems.

Before concentrating on the experimental situation - about which I will have much to say-I very briefly review some of the significant theoretical developments in the post Imry-Ma period.[4] Two basic objections appeared to the arguments given above leading to the conclusion $d_\ell = 2$ for the RFIM. The first involved the neglect of roughening of the wall between oppositely oriented domains and the second was the conflict with perturbation theory results. Though seemingly unrelated, these two considerations separately gave credence to the belief that $d_\ell = 3$ and had bearing on the erroneous interpretation that was given to early experimental work.

In the surface roughness argument the system is encouraged to gain still more random field energy by not requiring the interface to be smooth. Allowing the interface width w might grow in proportion to L lead some[5] to believe $d_\ell = 3$ at T = 0 K and that a roughening transition to a smoother interface state would occur at a finite $T = T_R$, which would restore the ordered state above d = 2.[6] These considerations appear to be incorrect in so far as it can be shown that, at T = 0 K, $w/L \sim h^{1/3}L^{(2-d)/3}$, which vanishes as $L \to \infty$ for $d > 2$.[4] Still more precise domain-wall arguments[7] have confirmed the original Imry-Ma picture and $d_\ell = 2$ for the RFIM.

The perturbation theory arguments start from a Ginzburg-Landau-Wilson representation of the RFIM and show that, order by order, there is an equivalence of the (d - 2)-dimensional pure problem to the d-dimensional random field problem.[8] Since the upper critical dimension d_u in the presence of the random field is 6, the theories represent ε expansions in ε = 6 - d. It had also been noted[9] that the hyperscaling relation $d\nu = 2-\alpha$ is violated for the random field problem, requiring instead (d-2) $\nu = 2-\alpha$ to order ε^3, (α and ν are the specific heat and correlation length exponents, respectively). This concept of a "dimensionality reduction" has been a thread woven through subsequent theoretical treatments even though it was thought the reduction might not be precisely 2 when far from d_u (e.g. at $d = 3$). Most germaine to the question of what is d_ℓ is, if the $d \rightarrow d - 2$ equivalence was exact at all d, it would follow that $d_\ell = 3$ for the RFIM because the pure Ising model has $d_\ell = 1$. Although still more sophisticated perturbative descriptions of the RFIM (*e.g.*, supersymmetry arguments),[10] also concluded $d_\ell = 3$, there has been increasing suspicion they are in error.[4] No rigorous refutation of the various perturbation theories has been given but virtually all interest in them seems to have evaporated following Imbrie's[11] proof, in late 1984, that $d_\ell = 2$ for the RFIM at $T = 0$ K.

Much of the theoretical effort expended on homing-in on d_ℓ was done at the expense of neglecting what the expected properties would be in the vicinity of the phase transition at $d > d_\ell$. Only in the last year or so has the attention of theorists shifted towards understanding the RFIM phase transition at $d > d_\ell$, both as to its static and dynamical properties. To be fair to the truly many theorists that have labored in the random field, they have been quick to see the errors in their ways. Almost without exception, they have leapt from the $d_\ell = 2$ to the $d_\ell = 3$ bandwagon and then back again, not looking over their shoulders nor licking their wounds. They have also been handicapped in making the decision to abandon the sinking $d_\ell = 3$ boat, by a set of mistakes that were made in the interpretation of early neutron scattering experiments.[12,13] Unlike the repentant theorists, these experimentalists have denied and obscured what are the simple experimental facts on the RFIM, beginning with their failure to disavow their "unambiguous" proof that $d_\ell \geq 3$!

More will be said later about scaling properties and recent theories and Monte Carlo simulations, of the dynamics of the RFIM, in the context of interpreting the experimental results. But now let us look at the experimental facts.

II. Experimental Facts

a. Setting the Stage

Let us suppose "ideal" realizations of $d = 2$ and $d = 3$ RFIM systems existed. What experiments would one perform to test the presumption that $d_\ell = 2$ and not 3? Two possibilities immediately suggest themselves; either look for the instability against domain formation and/or determine whether or not the phase transition is destroyed in a random field h. Both experiments are conceptually simple to do but how they are done - in particular, the route that is taken to reach a given h and T - turn out to have an important bearing on the final state in which the system finds itself. I begin with studies of the phase transition.

Rb_2CoF_4 and FeF_2 are prototypical Ising d = 2 and d = 3 uniaxial antiferromagnetic systems respectively. When substitutionally diluted with Mg^{2+} or Zn^{2+} ions ($Rb_2Co_xMg_{1-x}F_4$; $Fe_xZn_{1-x}F_2$) they become excellent examples of the corresponding d = 2 and d = 3 random exchange Ising model (REIM) systems. If a uniform field H is applied colinearly with the direction of sponaneous ordering, the respective diluted antiferromagnets behave as truly ideal d = 2 and d = 3 RFIM systems.[2] It was experimental studies of the phase transitions in these crystals which provided the critical test of the $d_\ell = 2$ prediction.

A caveat must be offered before discussing the phase transition studies of these or, for that matter, any randomly mixed system. In growing mixed crystals from the melt, a gradient of concentration of the two constituents is invariably present. For such systems (*e.g.* $Fe_xZn_{1-x}F_2$) the existence of a gradient translates directly into a spread in transition temperatures through the crystal. Some means must be had for determining the nature of the gradient, preferably in a way that is independent of the critical phenomena itself.

Fortunately, for the $Fe_xZn_{1-x}F_2$ system just such a means exists in measurements of the optical birefringence Δn at ambient temperature. The values of Δn in pure FeF_2 and ZnF_2 differ by 2% and, in the alloys $Fe_xZn_{1-x}F_2$, Δn varies monotonically between the two extremes. Thus, once having measured Δn *vs* x, one can determine the variation in x within a given crystal because Δn can be measured with a precision of 1 part in 10^8! Extremely small variations from point to point in a crystal may be probed using a laser beam plus pinhole collimator (of spot size width $\ell <$ of order 100 μ). An example of this is shown in Fig. 1 for an $Fe_{0.6}Zn_{0.4}F_2$ crystal, where the laser beam was oriented perpendicular to the scan direction as is shown in the inset.[14] Although the average gradient along the growth (z) direction is only 0.65%/cm, it causes a variation in T_N of $\delta T_N \simeq 0.35$ K across the 7mm crystal.

Because of the concentration gradient problem, the birefringence technique also

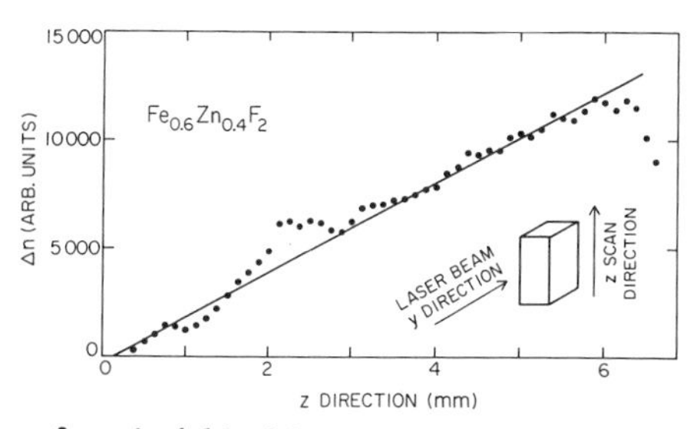

Fig.1. Variation of optical birefringence Δn at ambient temperature along the growth (z) axis. The variation in δn over the 7 mm length corresponds to concentration differential of $\delta x = 0.45$ mol%. The point-to-point variations are real fluctuations in concentrations - not noise!

turns out to be the best way for getting at the thermal critical exponent α associated with the magnetic specific heat divergence $C_m \sim |t|^{-\alpha}$. Whether H $=$ 0 or H $\neq$ 0, one can show that in the critical region $d(\Delta n)/dT \sim C_m$. This obtains because the most singular terms in all second derivatives of the free energy F ($\partial^2 F/\partial T^2, \partial^2 F/\partial T \partial H$ and $\partial^2 F/\partial H^2$) diverge as $H^y \mid t \mid^{-\alpha}$, with the same α but different field scaling exponents y,[15] and the fact that $d(\Delta n/dT)$ is a linear combination of $\partial^2 F/\partial T^2$ and $\partial^2 F/\partial T \partial H$. Because the laser beam can be directed perpendicular to the gradient,[16] the effects of the latter on rounding of the transition can be drastically reduced from what it would be if the whole crystal were used in a conventional C_m experiment. Precisely this was done in the two crucial measurements near the phase transitions in the prototypical d $=$ 2 and d $=$ 3 RFIM systems to be discussed next.

b) Is $d_\ell = 2$ or 3?

The critical region of the d $=$ 2 ($Rb_2Co_{0.85}Mg_{0.15}F_4$) and d $=$ 3 ($Fe_{0.6}Zn_{0.4}F_2$) systems was studied using the birefringence technique. In Fig. 2, $d(\Delta n)/dT$ *vs* T at H $=$ 0 and H $=$ 20 kOe (H$\|$c axis) is shown for the d $=$ 2 case, from which it is seen the transition is destroyed in the presence of a random field.[17] This is to be contrasted with the d $=$ 3 case of Fig. 3, where $d(\Delta n)/dT$ *vs* T is shown at H $=$ 0, 14 and 20 kOe (H$\|$c axis). Remarkably, the transition seems to sharpen with the application of a field.[16] Clearly the presence of a random field does not destroy the transition in the latter instance. From these two measurements alone one can conclude.[16,17]

$$2 \leq d_\ell < 3 \qquad (1)$$

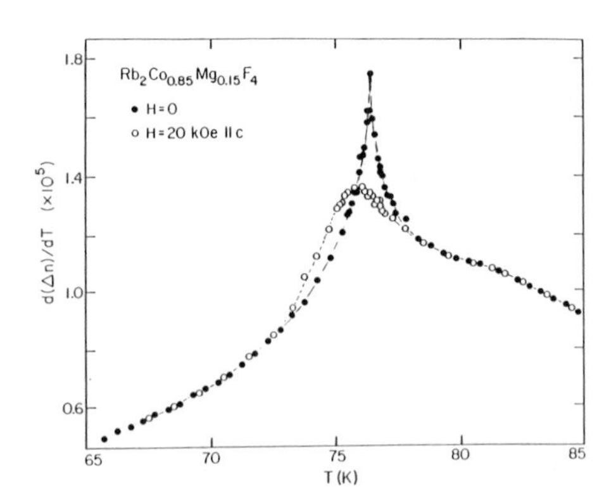

Fig. 2. d $d(\Delta n)/dT$ *vs.* T of the d = 2 RFIM system in applied fields H = 0 and H = 20kOe. The rounding of the peak corresponds to the destruction of the phase transition; ergo $d_\ell \geq 2$ for the RFIM.

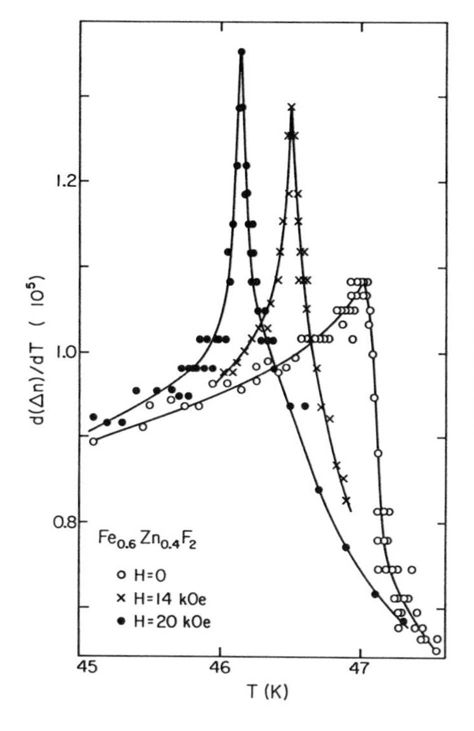

Fig. 3. $d(\Delta n)/dT$ *vs.* T of the d = 3 RFIM system in applied fields H = 0, 14 and 20 kOe. The absence of field induced rounding of the peak (in fact it clearly sharpens) implies no destruction of the phase transition; ergo $d_\ell < 3$ for the RFIM.

The importance of this result lies in the fact that all other measurements must be interpretable in terms of there being a sharp phase transition at d = 3 and a destroyed one at d = 2.

c) Hysteresis and the Dynamics of the RFIM

Knowing that $d_\ell < 3$, one might surmise the behavior of a d = 3 RFIM system in the vicinity of the phase transition $T_c(H)$ would be similar to a pure Ising antiferromagnet at $T_N(H)$. Nothing could be further from the truth! At d = 3, crossover from the random exchange to the random field fixed point results in entirely new critical behavior, which we will come back to later on. But from the experimental point of view what appears to be most distressing, at first sight, is that what one finds depends on the field-temperature route that one follows to arrive at a given (H,T). This is graphically illustrated by capacitance (C) studies on the d = 3 RFIM system as is shown in Fig. 4. Here $1/C(dC/dT)$ *vs* T is shown for H = 0 and H = 19 kOe in $Fe_{0.46}Zn_{0.54}F_2$ for two H-T routes to be described below.[18] This appearance of hysteresis is endemic to the RFIM problem and bears further discussion.

Since d_ℓ <3, cooling a randomly diluted, d = 3 Ising antiferromagnet (AF) in a field - so-called "field cooling" (FC) - should result in the onset of AF LRO

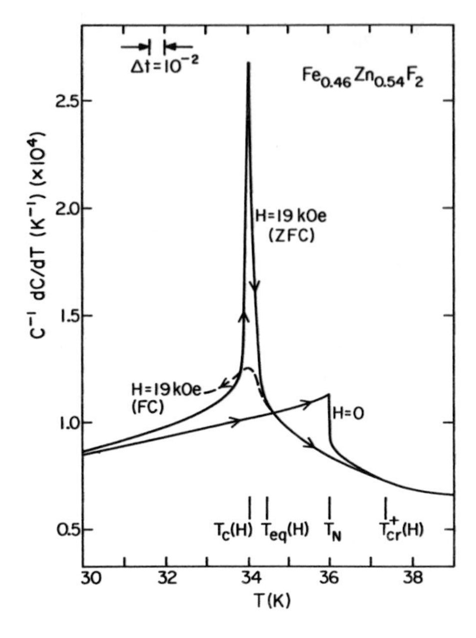

Fig. 4. $C^{-1}dC/dT$ *vs.* T of the d = 3 RFIM system in applied fields H = 0 and H = 19 kOe. The differences between the field-cooled (FC) and zero-field-cooled (ZFC) routes are all too apparent. The temperatures of the RFIM transition $T_c(H)$, the point above $T_c(H)$ where FC and ZFC first coincide, $T_{eq}(H)$, and the crossover to REIM behavior above $T_c(H)$, $T^+_{cr}(H)$ are as indicated. All obey crossover scaling (*i.e.* $\sim H^{2/\phi}$).

at $T_c(H)$. This is not what happens. Instead, at a temperature $T_{eq}(H)$ (defined below), which is slightly above $T_c(H)$, the system becomes trapped into a domain-like configuration from which it cannot extricate itself as T is lowered below $T_c(H)$, on what are laboratory time scales. However, if the system is cooled in zero field below $T_c(H)$ and then the field is applied - so called "zero field cooling" (ZFC) - the AF LRO persists for all values of H and T below $T_c(H)$. Clearly, if $d_\ell < 3$, the FC route does not bring the system to its true equilibrium state (*i.e.* one with LRO) below $T_c(H)$. $T_{eq}(H)$ is the temperature above which birefringence, capacitance and neutron scattering yield identical results whether FC or ZFC.

For the FC route to differ from the ZFC, there must exist extremely long equilibration times in the vicinity of $T_c(H)$. This appears to be a direct consequence of the unusual critical slowing down that occurs as $T_c(H)$ is approached.[19,20,21] It has been argued that the static random field fluctuations dominate over the thermal ones in the vicinity of $T_c(H)$. Consequently the characteristic time τ for the spin correlations to grow to a length ξ may happen extremely slowly, possibly as[19]

$$\tau \sim exp(C \mid t \mid^{-\nu\theta}) \tag{2}$$

where ν is the correlation length exponent ($\xi = \xi_o \mid t \mid^{-\nu}$) and $1 \leq \theta \leq 2$. Implicit in (2), or alternatively $\log \tau \sim \xi^\theta$, is that changes in H or T on time scales of order τ will result in the system being trapped into static domain configurations of dimension L that would scale as $L/a_o \sim \log(\tau/\tau_o)$, where a_o is the lattice spacing and τ_o is presumably an inverse exchange frequency. Before becoming enamored with a particular dynamic scaling relation, one might well ask if there is any experimental evidence for extreme slowing down in the critical dynamics. The answer to this is to be found in very recent studies of the uniform ac susceptibility $\chi(\omega)$.[22]

As remarked earlier, the leading RFIM singularity of $\partial^2 F/\partial H^2$ and $\partial^2 F^2/\partial T^2$ should have identical thermal exponents; hence $\chi \sim C_m \sim \mid t \mid^{-\alpha}$. From birefringence studies the latter has been observed to be a symmetric, logarithmic (*i.e.* $\alpha \simeq 0$) divergence.[16] Measurements of the ω dependence of the real part of the uniform ac susceptibility $\chi'(\omega)$ on the very homogeneously random crystal $Fe_{0.46}Zn_{0.54}F_2$ reveal two interesting features; the amplitude decreases and the effective width of the symmetric peak broadens with increasing ω at extremely low frequencies ($2\text{Hz} \leq \omega/2\pi \leq 8.5\text{kHz}$). This is shown in Fig. 5 for three frequencies at 10 kOe.

One may understand these findings as follows. The departure of $\chi'(\omega)$ from log $\mid t \mid$ behavior as $\mid t \mid \rightarrow 0$, after subtraction of an effective $\chi'(\infty)$ background, is to be associated with a characteristic dynamic rounding temperature $t^*(\omega)$, which is indicated by the arrows in Fig. 6. For temperatures $t > t^*(\omega)$ the system can equilibrate on time scales $\tau < 1/\omega$ but for $t < t^*(\omega)$ it cannot.

The ω dependence of $t^*(\omega)$ and the peak height $[\chi'(\omega)]_p$ can be interpreted in terms of conventional dynamic scaling, in which case $t^*(\omega) \sim \omega^{1/z\nu}$, with z the dynamic critical exponent. Because the dynamics limit the static divergence of $\chi'(\omega)$, scaling requires $[\chi'(\omega)]_p \sim \omega^{-\alpha/z\nu}$, which as $\alpha \rightarrow 0$ behaves as $\ell n\ \omega$. The behavior of $t^*(\omega)$ and $[\chi'(\omega)]_p$ *vs* ω are shown in Fig. 7 and can be well fit with $z \simeq 14$ - an extraordinarily large number as values of z go (usually $1 \leq z \leq 2$).

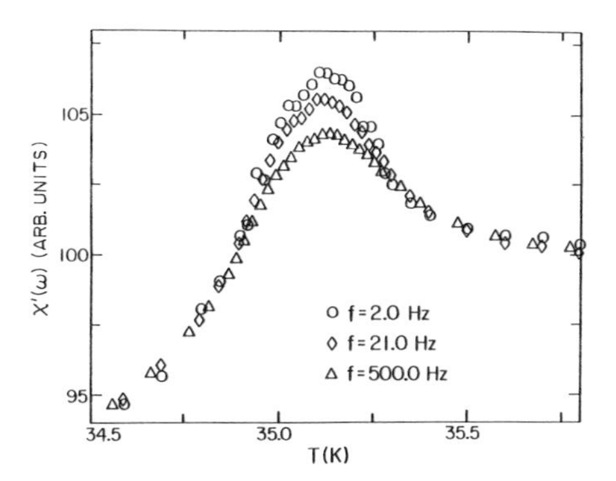

Fig. 5. Real part of ac susceptiblity $\chi'(\omega)$ *vs.* T near $T_c(H)$ of d = 3 RFIM system $Fe_{0.46}Zn_{0.54}F_2$ at H = 10 kOe, with H||c-axis, at three different frequencies.

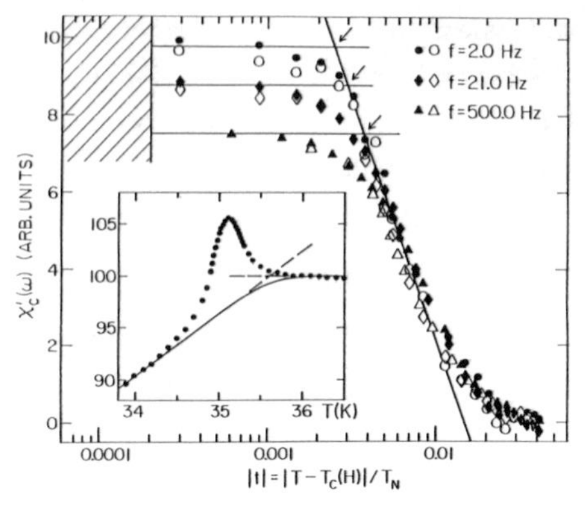

Fig. 6. $\chi\prime(\omega)$ *vs.* log $| t |$, the reduced temperature at same field and frequencies of Fig. 5. (Note $\chi_c(\omega) = \chi\prime(\omega)$ - $\chi\prime(\infty)$, with $\chi\prime(\infty)$ estimated as shown in inset by solid line). The open and closed circles refer to T < T_c(H) and T > T_c(H), respectively. Rounding of T_c(H) because of concentration gradient is confined to shaded area; *i.e.* $| t | < 2\times10^{-4}$. The expected log $| t |$ behavior to $\chi\prime(\omega)$ is used to determine a dynamic rounding temperature $t^*(\omega)$ at each ω, as indicated by arrows.

Using the "activated" dynamic scaling associated[19,20] with Eq. 2 results in $t^*(\omega) \sim (\ell n\ \omega)^{-1/\theta\nu}$ and $[\chi\prime(\omega)]_p \sim (\ell n\ \omega)^{\alpha/\nu\theta}$ which goes to $\ell n(\ell n\ \omega)$ as $\alpha \to 0$.

The frequency range of these measurements has been extended using the Faraday rotation method to a measurement time scale $\tau = 1/\omega \sim 100$ sec.[15b] Over the combined six decades in ω, either the conventional or "activated" dynamic scaling forms agree with experiment. Whether or not either of these scaling forms is the correct one, it is undeniable that the slowing down of d = 3 RFIM systems is so extreme that equilibration is virtually impossible for experimentally realizable times ($\tau < 10^4$ sec) below $| t | \simeq 10^{-3}$. Interestingly, $t_{eq} \equiv$

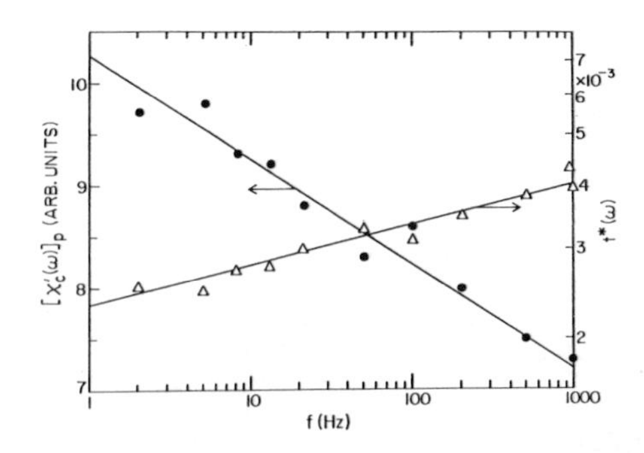

Fig. 7. Peak amplitude $[\chi\prime_c(\omega)]$ of $\chi\prime(\omega)$ and the logarithm of the dynamic rounding temperature t^* (ω) *vs.* log ω. (See text for conventional dynamic scaling interpretation.)

$[T_{eq}(H) - T_c(H)]/T_N > t^*(\omega, H)$ at the same H and τ, leading one to believe that the characteristic times for small amplitude fluctuations are very much shorter than are those that characterize domain wall motion at the same t and H.

There are two important consequences of this equilibration time limit. First, there is no way to access the LRO AF ground state by FC through the transition region and second, asymptotic RFIM critical experiments may not be determinable by experiment. The former consideration is the root cause of the observed hysteresis, insofar as a domain state configuration becomes frozen-in when FC through the transition region. Once slighly below T_c(H) the spontaneous sublattice magnetization M_s(T) rises very rapidly and the energy barriers become too large for thermal excitations to anneal out the higher energy, metastable domain configurations on laboratory time scales.

d) Crossover Scaling and the RFIM Critical Behavior

As pointed out in Section IIa, the diluted Ising AF become excellent examples of the REIM at H = 0. Mössbauer Effect studies[23] of the magnetization exponent β *vs* x in the $Fe_xZn_{1-x}F_2$ system indicate the crossover from pure Ising to REIM behavior is virtually complete for x < 0.9. Since everything to be discussed below is in this limit for both the d = 2 and d = 3 diluted AF, the expected crossover in a field is from REIM to RFIM and not pure Ising to RFIM behavior. The critical exponents for the d = 3 REIM system have recently been measured on the "super" crystal $Fe_{0.46}Zn_{0.54}F_2$ from a combined neutron scattering[24] Mössbauer Effect[23] and birefringence studies.[25] The values are given in Table I and are seen to be in very good agreement with renormalization group predictions[26] as well as satisfying the hyperscaling relation $2-\alpha = \nu d$.

In the presence of a random field h the free energy F has a scaling form

$$F = \mid t \mid^{2-\alpha} f(th^{-2/\phi}) \tag{3}$$

which exhibits the leading $\mid t \mid^{2-\alpha}$ behavior for h = 0. The quantity ϕ is the exponent which governs the crossover from REIM to RFIM critical behavior. It has just been shown by Aharony[27] that ϕ should not be identical to the REIM susceptibility critical exponent γ, as would be the case for nonrandom systems. Indeed, precision measurements[24,25] of γ and ϕ on the same d = 3 "super" crystal shows $\phi = 1.42 \pm 0.03$, yielding a ratio $\phi/\gamma = 1.08 \pm 0.05$ in agreement with Aharony's prediction of $1.05 \leq \phi/\gamma \leq 1.1$.

The importance of establishing the value of ϕ lies in the fact that all of thermodynamic functions (*e.g.* C_m), and all of the "phase boundaries" have their field

TABLE 1. CRITICAL EXPONENTS OF D=3 REIM AND RFIM SYSTEMS*

	RANDOM EXCHANGE		RANDOM FIELD	
	Exp't	Theory$^{(e)}$	Exp't$^{(f)}$	Theory$^{(g)}$
α	-0.09± 003$^{(a)}$	(-0.04)-0.09	0.00± 0.03	
ν	0.69± 0.01$^{(b)}$	(0.68)-0.70	1.0 ± 0.15	1.1± 0.05
γ	1.31± 0.03$^{(b)}$	(1.34)-1.39	1.75± 0.20	
β	0.36± 0.01$^{(c)}$	(0.35)0.35		<0.1
η	0.07± 0.04$^{(d)}$	(0.02)$^{(d)}$0.01$^{(d)}$	0.25± 0.36$^{(d)}$	0.5± 0.1

*All measurements on the diluted Ising AF $Fe_xZn_{1-x}F_2$ (a) Ref 35; (b) Ref 24; (c) Ref 23; (d) calculated from $\eta = 2\nu - \gamma$; (e) Ref 26a and (Ref 26b); (f) Ref 14; (g) Monte Carlo Simulations of Ref 21.

dependencies governed by crossover scaling. For example the phase transition temperature $T_c(H)$ is given by

$$T_c(H) = T_N - bH^2 - cH^{2/\phi} \tag{4}$$

where the bH^2 term is a mean-field (nonrandom dilution) correction and use has been made of the proportionality between h and H. Because of the explicit $H^{2/\phi}$ dependence, novel "phase diagrams" can be constructed[28] for both the d = 2 and d = 3 RFIM systems and are shown in Fig. 8 and Fig. 9. They become a convenient way of displaying the profound differences between RFIM systems for d at (d = 2), or above (d = 3), the lower critical dimension d_ℓ.

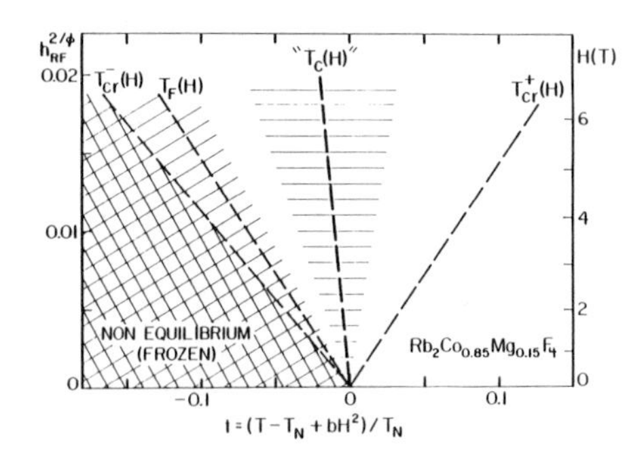

Fig. 8. New d = 2 RFIM "phase diagram". Shown are the scaling behavior of the location of (1) $T_F(H)$ and width slanted shading of the metastability boundary; $T_c(H)$ and width (horizontal shading) of the <u>destroyed</u> phase transition; and (3) REIM-RFIM crossover boundaries $T_{cr}^+(H)$ and $T_c^-(H)$. Crossover exponent ϕ =1.75.

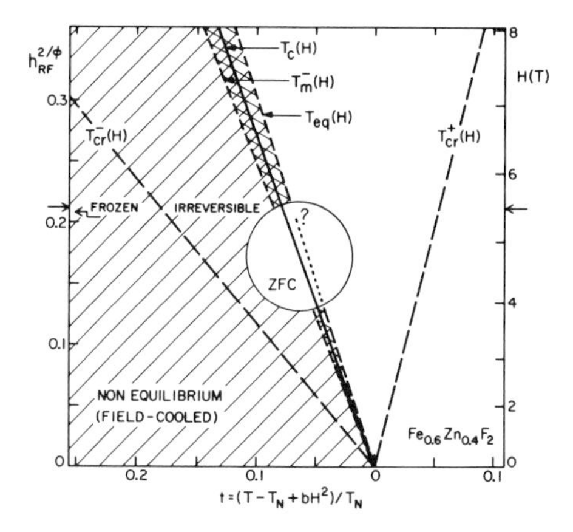

Fig. 9. New d = 3 RFIM "phase diagram". Shown are the scaling behavior of the location of (1) sharp phase transition at $T_c(H)$; (2) FC equilibrium boundary $T_{eq}(H)$; REIM-RFIM crossover boundaries $T_{cr}^+(H)$ and $T_{cr}^-(H)$. If FC a nonequilibrium domain configuration exists in entire region below $T_{eq}(H)$ and metastable behavior is exhibited between $T_m^-(H)$ and $T_c(H)$. If ZFC, no irreversible or metastable behavior is found anywhere below $T_c(H)$. Crossover exponent ϕ =1.40.

The "phase diagram" for the d = 2 $Rb_2Co_{0.85}Mg_{0.15}F_4$ system shows the position of the rounded peak in C_m, which characterizes the destroyed phase transition at "$T_c(H)$ ", and its width (horizontally shaded region). Also seen are the "crossover" boundaries as obtained from measurements of $d\Delta n)/dT$ *vs* T as well as the metastability boundary $T_F(H)$ and its width (indicated by the single slanted shading). $T_F(H)$ was determined via a neutron scattering experiment in which the system was first ZFC to well below "$T_c(H)$", followed by the field being raised to a particular value H and then slowly warmed. Since the ZFC (LRO) state is not the ground state of the d = 2 RFIM the system evolves into a domain state as it unfreezes. On a given time scale this decay towards an equilibrium domain configuration occurs over a relatively narrow region of T.[29] In the vicinity of $T_F(H)$ the metastable ZFC configuration is observed to relax logarithmically with time as is shown in Fig. 10. (all of this was missed in the early neutron scattering experiments[12,13] because only FC was done!). The logarithmic time dependence for the evolution of metastable RFIM configurations had been surmised in earlier theoretical work.[30] It should be remarked that neither ZFC nor FC procedures allow one to access the Imry-Ma ground state on laboratory time scales at T = 0K because the system freezes at $T \simeq T_F(H)$. Hence, even if FC, the field dependence of the domain size at low T is not characteristic of the Imry-Ma d = d_ℓ ground state at d = d_ℓ [i.e. $L \sim \exp h^{-2}$]. Instead the size is determined by what is the equilibrium domain configuration at $T = T_F(H)$.

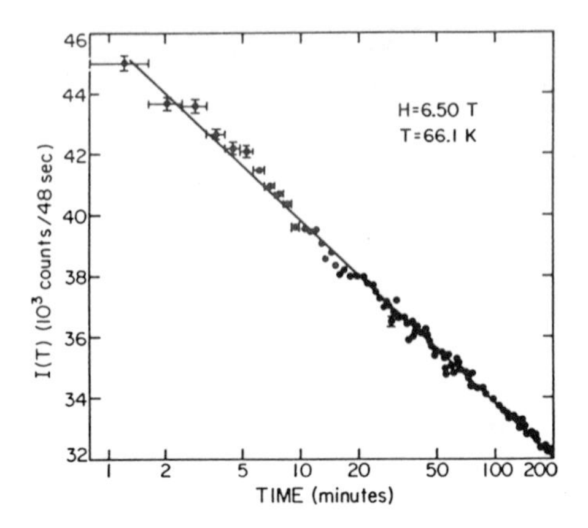

Fig.10. Metastability in d=2 RFIM system $Rb_2Co_{0.85}Mg_{0.15}F_4$ at $T_F(H)$ (see Fig. 8). Shown is intensity I(T) of ZFC Bragg peak *vs.* time in semilog plot at field of H=65kOe.

By way of contrast, the d = 3 "phase diagram" for the $Fe_{0.6}Zn_{0.4}F_2$ system shows only the sharp phase boundary at $T_c(H)$ and the crossover boundaries [$d(\Delta n)/dT$ or dC/dT *vs* T determined] if ZFC. However, if FC, the system becomes trapped into a nonequilibrium domain configuration at $T_{eq}(H) > T_c(H)$, with $T_{eq}(H) - T_c(H) \sim H^{2/\phi}$. The scaling of this equilibrium boundary was first established with capacitance studies (Figs. 4 and 11) but has subsequently been confirmed in $d(\Delta n)/dT$[25] and neutron scattering studies[31] in the "super" crystal $Fe_{0.46}Zn_{0.54}F_2$. Time dependent (metastable) behavior was observed between $T_m^-(H)$ and $T_c(H)$ if the system was field cooled. Again the time dependence appears to be logarithmic. (See Fig. 12).

The differences between the $d = d_\ell = 2$ and $d = 3 > d_\ell$ RFIM systems are striking. In the former the boundary that signals the onset of equilibrium with respect to domain formation lies well below the destroyed transition "$T_c(H)$" whereas in the latter the corresponding boundary is above the sharp phase transition at $T_c(H)$. It is clear that all "freezing" or "metastability" boundaries (whether at d = 2 or d = 3) must be defined on a given experimental time scale. Nevertheless, if that time scale is prescribed, the field dependence follows crossover scaling. What surely is not temporal is the continuous phase transition at $T_c(H)$ for the d = 3 RFIM system, contrary to the conclusions of Birgeneau *et al.*[32]

For the d = 3 system well defined critical behavior, characteristic of the RFIM fixed point, should be in evidence for $|t| < h_{cr}^{2/\phi}$, with the crossover boundaries indicated in Fig. 9. A lower bound in reduced temperature $|t| > t^*(1/\tau)$ is prescribed by the unusual dynamics of the RFIM, as discussed earlier. Here τ is

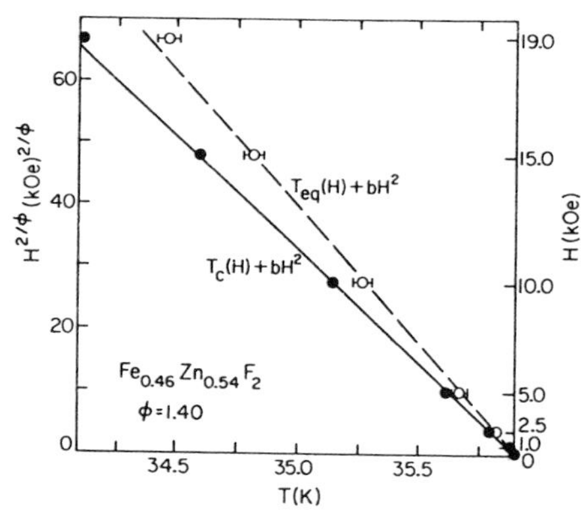

Fig.11. Capacitance determination of $T_c(H)$ and $T_{eq}(H)$ in the d=3 RFIM system. Both are plotted linearly *vs.* $H^{2/\phi}$, after meanfield nonrandom corrections. Only above $T_{eq}(H)$ are the FC and ZFC results for dC/dT identical.

the characteristic measurement time, realistically bounded to be no larger than 10^4 sec.

From combined birefringence[16] and neutron scattering studies[14] on the $Fe_{0.6}Zn_{0.4}F_2$ system the following values have been obtained for the d = 3 RFIM specific heat ($\bar{\alpha}$), correlation length ($\bar{\nu}$) and staggered susceptibility ($\bar{\gamma}$) critical exponents: $\bar{\alpha} = 0.00\pm0.03$, $\bar{\nu} = 1.0\pm0.15$, $\bar{\gamma} = 1.75\pm0.20$ and the correlation function exponent $\bar{\eta} \simeq 1/4$. The latter quantity is only indirectly determined through the assumed Lorentzian plus Lorentzian-squared form to the neutron scattering line shape and is subject to the most uncertainty.[14] In any case, the values of $\bar{\alpha}, \bar{\nu}, \bar{\gamma}$

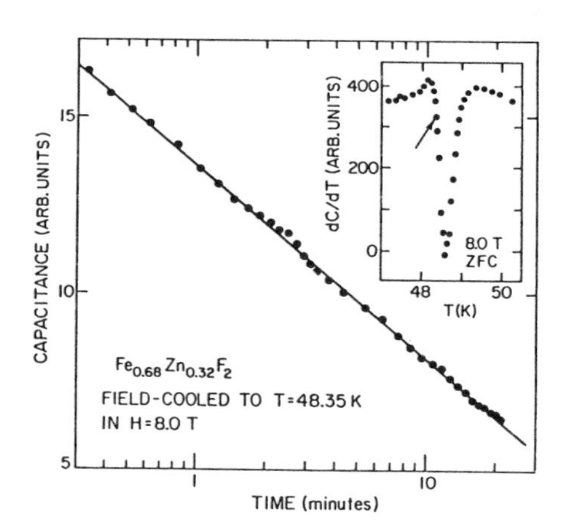

Fig.12. C *vs.* logarithmic of time for d=3, nonequilibrium FC RFIM system. The inset shows the ZFC results for dC/dT and the arrow indicates the value of T at which the time dependent measurements of C were made following the FC procedure.

and $\bar{\eta}$ appear to be close to those for a pure d = 2 Ising system, leading one to believe there may exist a dimensionality reduction $\bar{d} \simeq$ d-1 for the d = 3 RFIM system. The RFIM and REIM exponents are collected in Table 1. A more detailed recent review of the experimental work of the UCSB-UCSC collaboration has been given by King and Belanger.[33]

III. Experimental "Fiction"

a) Unambiguous Demonstration of d_ℓ

One of the first RFIM experiments problem was a neutron scattering study of the d = 2 and d = 3 randomly diluted AF $Rb_2Co_{0.7}Mg_{0.3}F_4$ and $Co_{0.3}Zn_{0.7}F_2$.[12] Briefly stated, the experiments consisted of measurements of the structure factor $S(\vec{q})$ associated with scattering near the (100) AF Bragg peak. Only FC studies were made. It was found that the widths of the peak at low T increased with increasing H, indicating AF long range order was destroyed in both the d = 2 and d = 3 systems. The authors concluded:[12] "This simple result alone demonstrates unambiguously that $d_\ell \geq 3$".

Taking the structure factor to have the form $S(q) = A(\kappa^2+q^2)^{-2}+B(\kappa^2+q^2)^{-1}$, they interpreted κ to be an inverse "correlation length" and found a power law field dependence to κ ($H^{1.6}$) for the d = 2 system. Since extant theories[34] predicted the size of the Imry-Ma domains at $d = d_\ell$ should scale as $\kappa \sim exp(-cH^{-2})$, whereas at $d < d_\ell$ $\kappa \sim H^2$, the authors concluded: "These results therefore necessitate $d_\ell >$ 2 and are consistent with $d_\ell = 3$."

These conclusions we know now to be incorrect. The FC procedure cannot access the ground state of the RFIM at d = 3. At d = 2 the FC procedure will result in some domain configuration being frozen-in at $T_F(H)$, as we discussed earlier. Unfortunately, this is not the equilibrium domain configuration that was predicted in the theory at T = 0K and which led to the $\kappa \sim exp(-CH^{-2})$ scaling with H at d = d_ℓ . In a later more detailed report on the d = 2 $Rb_2Co_{0.7}Mg_{0.3}F_4$ RFIM system, Birgeneau *et al*[13] argued that the absence of equilibration at low temperatures "-cannot vitiate our principal conclusion that $d_\ell > 2$". To my knowledge, Birgeneau and collaborators have yet to repudiate the interpretation given in their earlier work.

b) The Temporal Phase Transition at d = 3

In what appears to be a complete turnabout, Birgeneau *et al* have gone from believing there was no sharp, continuous phase transition at $T_c(H)$ to the opposite extreme of a discontinuous one at d = 3. Their new work[32] appeared under the title "Temporal Phase Transition in Three-Dimension Random-Field Ising Model".

(Nowhere in the abstract or the body of the paper are the words "temporal phase transition" used again or any hint of a definition given.) The substance to this latest study on $Mn_{0.75}Zn_{0.25}F_2$ is the claim that upon FC there is "a well-defined transition to a state with extremely long relaxation times and many metastable states". It is also asserted "–the loss of LRO occurs discontinuously and that above this boundary the sample is always in equilibrium."

I submit that Birgeneau *et al* have erred once again insofar as they have used a crystal with a concentration gradient that is large enough to have obscured the differences between $T_c(H)$ and $T_{eq}(H)$ (see Fig.11). In a preliminary neutron scattering study of the RFIM phase transition region in the "super" crystal $Fe_{0.46}Zn_{0.54}F_2$, we have found no evidence for a discontinous loss of LRO and that $T_{eq}(H)$ and $T_c(H)$ could be separately identified.[31] The FC determined $T_{eq}(H)$ does not correspond to any phase transition; "temporal", "first-order" or otherwise. Recall, no specific heat anomaly exists at $T_{eq}(H)$.

Indeed, LRO will disappear in a narrow temperature range, below $T_c(H)$, but not because there is a first-order transition, but rather as a consequence of the extremely small value of $\bar{\beta}$, the RFIM d = 3 magnetization exponent. Recent Monte Carlo simulations suggest that $\bar{\beta} < 0.1$ for the d = 3 RFIM.[21] To be sure then a measurement of $\bar{\beta}$ will require a crystal with an extremely small concentration gradient, which would certainly rule out the one used in the recent Birgeneau *et al*[32] experiment.

c) The Wages of Sin

I believe the failure to identify the gradient of concentration in a mixed crystal has caused Birgeneau *et al* to have erred in the determination of the d = 3 REIM susceptibility exponent γ and the crossover exponent ϕ in the dilute AF $Fe_{0.5}Zn_{0.5}F_2$ from neutron scattering experiments. They had found $\gamma = 1.44 \pm 0.06$[35] and $\phi = 1.5 \pm 0.15$.[36] After their experiments were completed the crystal was examined in our laboratories and the concentration gradient measured.[37] The part of the crystal used for the neutron scattering studies had a variation in concentration such that $\delta T_N/T_N = 1.4\times10^{-2}$, whereas Birgeneau *et al* had reported critical scattering measurements down to $|t| = 10^{-3}$! Not knowing the concentration gradient in the $Mn_{0.75}Zn_{0.25}F_2$ sample has resulted in Birgeneau *et al*[32] finding $\bar{\nu} = 1.52 \pm 0.13$ in conflict with Monte Carlo simulations[21] which give $\bar{\nu} = 1.1 \pm 0.05$ and our value of $\bar{\nu} = 1.0 \pm 0.15$.

IV. What Remains to be Understood?

Accepting the premise that $d_\ell = 2$ and the properties of the d = 3 RFIM system are largely governed by the unusual dynamic critical behavior at the continuous phase transition, there still remains a number of questions to be answered. Some that are of interest to the author and his colleagues are the following:

1. Can one measure the asymptotic static critical behavior of the d = 3 RFIM problem despite the extreme slowing down of the critical fluctuations?

2. Do the d = 3 RFIM fixed point exponents correspond to some pure Ising (possibly non-integral) reduced dimensionality or is dimensionality reduction now to be discarded as a theoretical fiction?

3. Are there other critical points along the d = 3 RFIM phase boundary at large values of h?

4. What are the appropriate forms for S(q) for both the d = 2 and d = 3 REIM and RFIM systems?

5. Are there any essential differences in the scaling and critical properties of weak (*e.g.* $Mn_xZn_{1-x}F_2$) and strong (*e.g.* $Fe_xZn_{1-x}F_2$) anisotropy, randomly diluted AF, either in zero or a finite field?

6. Is there a lesson to be learned from the happenings of the random field problem? Four centuries ago it was observed:

> "Once people are wedded to a certain idea or concept, their minds are no longer open. They will continue to look for evidence, however flimsy it may be, to buttress preconceived notions, and will reject all evidence that contradicts them".

Francis Bacon (1561-1626).

Acknowledgements

To be able to "see" the distant truths one has to stand on the shoulders of able colleagues. D. P. Belanger and A. R. King have been the mainstays of our experimental effort and J. L. Cardy has provided theoretical guidance and insights. I. B. Ferreira, J. Mydosh, W. Kleemann, and S. M. Rezende, have been active participants in our UCSB research program. J. Arthur, R. Nicklow and O. A. Pringle have given themselves selflessly in our collaborative neutron scattering studies at Oak Ridge National Laboratory. A. T. Ogielski made me aware of the significance of his Monte Carlo simulation results. Finally, my thanks to A. Aharony whose 1981 lecture at the ITP first brought the RFIM problem to our attention. The research at UCSB has been supported by NSF DMR Grant Numbers 80-17582 and DMR85-16786.

1. Y. Imry and S.-k Ma, *Phys. Rev. Lett.* **35**, 1399 (1975).

2. S. Fishman and A. Aharony, *J. Phys* **C12**, L729 (1979).

3. J. L. Cardy, *Phys. Rev. B* **29**, 505 (1984).

4. A good review of the pre-1984 state of the theory is given in G. Grinstein and S.-k Ma, *Phys. Rev. B* **28**, 2588 (1983).

5. See for example, D. Mukamel and E. Pytte, *Phys. Rev. B* **25**, 4779 (1982).

6. S. T. Chui and J. D. Weeks, *Phys. Rev. B* **14**, 4978 (1976).

7. J. Villain, *J. de Phys. (Paris)* **43**, L551 (1982) and ref. 4.

8. A. Aharony, Y. Imry and S.-k. Ma, *Phys. Rev. Lett.* **37**, 1364 (1976).

9. G. Grinstein, *Phys. Rev. Lett.* **37**, 944 (1976).

10. G. Parisi and N. Sourlas, *Phys. Rev. Lett.* **43**, 744 (1979).

11. J. Z. Imbrie *Phys. Rev. Lett.* **53**, 1747 (1984).

12. H. Yoshizawa, R. A. Cowley, G. Shirane, R. J.Birgeneau, H. J. Guggenheim and H. Ikeda, *Phys. Rev. Lett.* **48**, 438 (1982).

13. R. J. Birgeneau, H. Yoshizawa, R. A. Cowley, G.Shirane and H. Ikeda, *Phys. Rev. B* **28**, 1438 (1983).

14. D. P. Belanger, A. R. King and V. Jaccarino, *Phys. Rev. B* **31**, 4538 (1985).

15. J. L. Cardy private communication and W. Kleemann, A. R. King and V. Jaccarino, *Phys. Rev. B Rapid Communications* (1986) to appear.

16. D. P. Belanger, A. R. King, V. Jaccarino and J. L. Cardy, *Phys. Rev. B* **28**, 2522 (1983).

17. I. B. Ferreira, A. R. King, V. Jaccarino, J. L. Cardy and H. J. Guggenheim, *Phys. Rev. B* **28**, 5192 (1983).

18. A. R. King, S. M. Rezende and V. Jaccarino unpublished .

19. J. Villain, *J. Phys. (Paris)* **46**, 1843 (1985).

20 D. S. Fisher, *Phys. Rev. Lett.* **56**, 416 (1986).

21. A. T. Ogielski and D. A. Huse, *Phys. Rev. Lett.* **56**, 1298 (1986); and private communication.

22. A. R. King, J. A. Mydosh and V. Jaccarino, *Phys. Rev. Lett.* (1986),

to appear.

23. P. Barrett (presented at ICM'85, San Francisco) and to be published.

24. D. P. Belanger, A. R. King and V. Jaccarino, *Phys. Rev. B* (1986), to appear.

25. I. B. Ferreira, A. R. King and V. Jaccarino, *Phys. Rev. B* (1986), to be submitted.

26. a) K. Newman and E. K. Reidel, *Phys. Rev. B* **25**, 264 (1982); b) G. Jug, *Phys. Rev. B* **27**, 609 (1983).

27. A. Aharony, *J. Europhysics Letters*, (1986) to be published.

28. A. R. King and V. Jaccarino, *J. Appl. Phys.*, **57** (1), 329, (1985).

29. D. P. Belanger, A. R. King and V. Jaccarino, *Phys. Rev. Lett.* **54**, 577 (1985).

30. J. Villain, *Phys. Rev. Lett.* **52**, 1543 (1984); R. Bruinsma and G. Aeppli, *Phys. Rev. Lett.* **52**, 1547 (1984); G. Grinstein and J. F. Fernandez, *Phys. Rev. B* **29**, 6839 (1984).

31. D. P. Belanger, A. R. King, V. Jaccarino, J. Arthur and R. Nicklow (1986) unpublished.

32. R. J. Birgeneau, R. A. Cowley, G. Shirane and H. Yoshizawa, *Phys. Rev. Lett.* **54**, 2147 (1985).

33. A. R. King and D. P. Belanger, *J. Mag. and Mag. Mat.*, **54-57**, 19, (1986).

34. A. Aharony and E. Pytte, *Phys. Rev. B* **27**, 5872 (1983).

35. R. J. Birgeneau, R. A. Cowley, G. Shirane, H. Yoshizawa, D. P. Belanger, A. R. King and V. Jaccarino, *Phys. Rev. B* **27**, 6747 (1983).

36. R. A. Cowley, H. Yoshizawa, G. Shirane and R. J. Birgeneau, *Z. Phys.* **B58**, 15 (1984).

37. I. B. Ferreira, A. R. King and V. Jaccarino, *unpublished.*

Density Functional Theory of Excited States

Walter Kohn

Preface.

I would like to begin with a few personal remarks about Ted, about how I remember Ted.

In 1950, I came to Pittsburgh after four years in Cambridge (Massachusetts). There I had done my graduate work and been a postdoc/instructor, while incredibly exciting things were happening around me: Quantum Electrodynamics, Nuclear Magnetic Resonance, Electron Spin Resonance, important new particles (π–mesons...), S-matrix Theory, etc. Those were heroic times and Cambridge was a heroic place.

Moving to Carnegie in Pittsburgh was, to be honest, a bit of a culture shock. Fred Seitz and his group had just moved to Illinois (one reason why I was hired) and Wick had not yet come. After the heady years in Cambridge it was not easy for me to get adjusted.

Fortunately, after a few weeks I met Ted, I am pretty sure at a Westinghouse seminar, where he was reigning royalty. I still vividly remember him in those surroundings: blunt-spoken, pointing right to the heart of the matter, re-defining issues in his own terms, very physical and anti-formalistic, judgmental (ranging from "wrong" and "trivial" to "fairly interesting") and self-critical. His absolute intellectual integrity and his total commitment to good physics were the two pillars on which his scientific personality rested. Those of you who were in Pittsburgh at the time will recall the awe with which most of his colleagues regarded him. I realized quickly how much I could learn from him, attitudes and insights, and felt much better about my new scientific home.

When Ted got into a problem, he stayed with it, 6 months, 1 year, 5 years... until he understood it to his satisfaction–which was not easy! In the meantime there appeared a stream of research reports, a kind of intellectual diary, moving forward and backward, self-correcting, raising new questions; and there were occasional seminars. For those of us near Ted, his working in our midst became an invaluable educational experience.

Ted and I always had the best of relations. Even though our work did not overlap much and our styles were rather different, Ted took an interest in what I was doing. In the early 60's I came to California and was very glad when, a few years later he came to UCLA. In the last few years I saw a little more of him (not enough) both in Santa Barbara and in Los Angeles. When I passed through a difficult personal period Ted gave me much-appreciated support.

I am very glad to dedicate this lecture to Ted. It isn't quite his cup of tea but I would like to hope that, nonetheless, he might have judged it as "fairly interesting."

1. Introduction.

Over the last 20 years density functional theory in the local density approximation (LDA) has been widely used for groundstate calculations.[1] (We recall here that in the LDA one approximates the true exchange correlation functional by

$$E_{xc}[n(r)] = \int e_{xc}(n(r))\,dr \qquad (inLDA), \tag{1}$$

where $e_{xc}(n)$ is the exchange correlation energy of a uniform electron gas of density n). The method has also been used for excited states[1] but, with the exception of a few special cases, without firm theoretical basis.

Soon after the original work of Hohenberg and Kohn (HK)[2] formal density functional theory was generalized to finite temperature ensembles by Mermin[3] and self-consistent equations, both for the ground state and for finite temperature ensembles where derived by Kohn and Sham (KS).[4]

In 1979 Theophilou[5] generalized the HK and KS theory to another ensemble, i.e. the mixture of the M lowest eigenstates equally weighted. We shall call this ensemble the equiensemble. The underlying reason for the possibility of generalizing the HKS theory to this ensemble is the following. The HK variational principle for the ground state energy E_1, has its origin in the Rayleigh-Ritz variational principle for E_1. Since there is also a generalized Rayleigh-Ritz principle for the quanity

$$< E_m >^M \equiv \frac{1}{M}(E_1 + E_2 + ...E_M) \tag{2}$$

an HK-like variational principle could be obtained for this quantity.

The results of the Theophilou theory are the following: There exists a functional $F^M[n(r)]$ of the mean density

$$n(r) \equiv < n_m(r) >^M \tag{3}$$

of the M-equiensemble, which is defined as

$$F^M[n(r)] \equiv < T + U >^M \quad _{n(r)}, \tag{4}$$

where T and U represent, respectively, the kinetic and interaction energy operators, and the subscript n(r) signifies the average over that equiensemble whose average density, Eq. (3), is $n(r)$. In terms of this functional $F^M[n(r)]$, one can construct the energy functional,

$$E_v^M[n(r)] \equiv \int v(r)n(r)\, dr + F^M[n(r)], \tag{5}$$

which is stationary with respect to $n(r)$ and whose minimum value is the true energy of the equi-ensemble. It is clear that calculating $E^{M=1} = E_1, E^{M=2} = \frac{1}{2}(E_1 + E_2)$ etc. the energies of successive excited states can be obtained.

Theophilou also generalized the ground state KS equations to the case of the equi-ensemble. They require a knowledge of the exchange correlation functional, $E_{xc}^M[n(r)]$, for the equiensemble, which is a generalization of the previously mentioned functional, $E_{xc}[n(r)]$, for the ground state. The form of the self-consistent equations is

$$\left[-\frac{1}{2}\nu^2 + v_{eff}(r)\right]\psi_i(r) = \epsilon_i\psi_i(r) \tag{6}$$

$$n_m(r) \equiv \sum |\psi_i(r)|^2 f_{im} \qquad\qquad f_{im} = 0 \ or \ 1 \tag{7}$$

$$n(r) \equiv < n_m(r) > M \tag{8}$$

$$v_{xc}(r) = \frac{\delta\, E_{xc}^M[n(r)]}{\delta\, n(r)} \tag{9}$$

$$v_{eff}(r) = v(r) + \int \frac{n(r')}{|r-r'|}\, dr' + v_{xc}(r) \tag{10}$$

Here $v_{eff}(r)$ is an effective trial potential; f_{im} is the occupation number of orbital i in the N-particle determinant describing the m'th excited non-interacting many body wave-function formed from the ψ_i; and $v(r)$ is the external single particle potential. The quantity E_{xc}^M is defined as

$$E_{xc}^M[n(r)] = F^M[n(r)] - \frac{1}{2}\int \frac{n(r)n(r')}{(r-r')}\, dr\, dr' - T_s^M[n(r)], \tag{11}$$

where

$$T_s^M[n(r)] \equiv kinetic\ energy\ of\ the\ non-interacting\ equiensemble\ (M; n(r)) \tag{12}$$

2. The Quasi-Local-Density Approximation for $E_{xc}^{M}[n(r)]$—No Trivial Answer.

In order to make practical use of the Theophilou theory, one requires a useful approximation for the key-functional $E_{xc}^{M}[n(r)]$. Since for the ground state, the local density approximation, Eq. (1), has been extremely useful one naturally looks for a similar approximation for the equiensemble.

$$E_{xc}^{M} = \int e_{xc}^{M}(n(r))\, dr \qquad (LDA) \tag{13}$$

where one hopes that, for each M, $e_{xc}^{M}(n)$ will be a universal function of n. Unfortunately this approach is not possible.

To see this impossibility consider, as an example, the uniform system for fixed n and M with periodic boundary conditions. n and M do not uniquely define the system. It also depends on the volume V (or the number of particles $N = Vn$). Now for a large systems, the levels will be very closely spaced and hence, for any finite M

$$e_{xc}^{M}(n) \to e_{xc}(n) \qquad (for\, V \to \infty), \tag{14}$$

where $e_{xc}(n)$ is the ground state exchange correlation energy. On the other hand for a small system, the levels are widely spaced and hence, for physical reasons,

$$e_{xc}^{M}(n) \neq e_{xc}(n). \tag{15}$$

Thus there is no universal function $e_{xc}^{M}(n)$ of the two variables n and M. How can one overcome this difficulty?

3. Use of Temperature and Thermodynamic Ensembles.

For the moment consider the case where M is sufficiently large. Then we know from quantum thermodynamics that:

$$Equiensemble(M, n(r)) \equiv Canonical\, ensemble\, (\theta, n(r)) \tag{16a}$$

$$with \qquad k\, \ell n\, M = S(\theta, [n(r)]); \tag{16b}$$

here θ is temperature and $S(\theta, [n(r)])$ is the (in principle known) entropy of a system with density n(r) and temperature θ. Equation (16b) defines the temperature θ.

Similarly, for a non-interacting (NI) equiensemble with density n(r) we have

$$NI\, equiensemble\, (M, n(r)) \equiv NI\, canonical\, ensemble\, (\theta_s, n(r)) \tag{17a}$$

$$with\; k\, \ell n\, M = S_s(\theta_s, [n(r)]) \tag{17b}$$

where $S_s(\theta', [n(r)])$ is the (in principle known) entropy of a NI system of density $n(r)$ and temperature θ'. Since the functions $S(\theta', [n(r)])$ and $S_s(\theta', [n(r)])$ occurring, respectively, in equations (16b) and (17b) are evidently different (one refers to an interacting, the other to a non-interacting system) the two temperatures θ and θ_s defined by Eqs. (16b) and (17b) are <u>different</u>.

In view of (16a) and (17a), the expression (11) for $E_{xc}^{M}[n(r)]$ can be written in terms of thermal averages.

$$E_{xc}^{M}[n(r)] = < T + V >_{n(r)}^{\theta} - \frac{1}{2} \int \frac{n(r)n(r')}{(r-r')} drdr' - T_s^{\theta_s}[n(r)], \qquad (18)$$

where θ and θ_s are given, respectively, by Eqs. (16b) and (17b), and

$$T_s^{\theta_s}[n(r)]$$
$$\equiv kinetic\, energy\, of\, the\, non-interacting\, canonical\, ensemble(\theta_s; n(r)) \qquad (19)$$

Eq. (18) should be compared to the exchange-correlation energy of the canonical ensemble

$$E_{xc}^{\theta}[n(r)] = < T + V >_{n(r)}^{\theta} - \frac{1}{2} \int \frac{n(r)n(r')}{|r-r'|} drdr' - T_s^{\theta}[n(r)], \qquad (20a)$$

so that we can write

$$E_{xc}^{M}[n(r)] = E_{xc}^{\theta}[n(r)] + \Big(T_s^{\theta}[n(r)] - T_s^{\theta_s}[n(r)]\Big). \qquad (20b)$$

4. <u>The Quasi-local-density Approximation.</u>

The expressions (20a) or (20b) for $E_{xc}^{M}[n(r)]$ are <u>exact</u>, for sufficiently large M, regardless of whether $n(r)$ is slowly or rapidly varying as function of r.

We now aassume that $n(r)$ is slowly varying. Then, in Eq. (20b) the <u>thermal</u> functionals, for <u>given</u> temperatures, can be written as spatial integrals of quantities depending only on local densities:

$$E_{xc}^{\theta}[n(r)] + T_s^{\theta}[n(r)] = \int \Big[e_{xc}^{\theta}(n(r)) + t_s^{\theta}(n(r))\Big] dr = \int e^{\theta}(n(r)) dr, \quad (21)$$

where $e_{xc}^{\theta}(n)$ and $t_s^{\theta}(n)$ are, respectively, the exchange-correlation energy of an uniform interacting electron gas and the kinetic energy of a uniform NI electron gas of density n; and

$$e^{\theta}(n) \equiv e_{xc}^{\theta}(n) + t_s^{\theta}(n). \qquad (22)$$

Similarly, the last term in Eq. (20b) becomes

$$T_s^{\theta_s}[n(r)] = \int t_s^{\theta_s}(n(r)) dr. \qquad (23)$$

Substituting Eq. (22) and (23) in Eq. (20b) gives finally

$$E_{xc}^{M}[n(r)] = \int \Big[e^{\theta}(n(r)) - t_s^{\theta_s}(n(r))\Big] dr. \qquad (24)$$

We recall that θ and θ_s are determined by M and $n(r)$ according to Eqs. (16b) and (17b). For slowly varying density, $n(r)$, these can be re-written as

$$k \log M = \int \sigma^{\theta}(n(r))dr \tag{25}$$

and

$$k \log M = \int \sigma_s^{\theta_s}(n(r))dr, \tag{26}$$

where $\sigma^{\theta'}(n)$ and $\sigma_s^{\theta'}(n)$ are, respectively, the entropy densities of a uniform interacting, and non-interacting, electron gas of density n. These relationships implicitly define θ and θ_s in terms of M and $n(r)$. Clearly, by (25) and (26) for a given M, θ and θ_s are totally non-local functionals of $n(r)$. However, their determination requires only a knowledge of the local functions $\sigma^{\theta}(n(r))$ and $\sigma_s^{\theta_s}(n(r))$. This is why we call the approximation (24), with θ and θ_s given by (25) and (26), a quasi-local-density approximation.

It is now easy to calculate $v_{xc}(r)$, Eq. (9), for a given M and $n(r)$ and thus to carry out the solution of the self-consistent KS equations ((6)-(10)). For details we refer the reader to a recent paper by the author.[6]

5. The Problem of Small M.

The quasi-local approximation of the previous two sections assumed a "sufficiently large" M. What does this mean?

Let us take a system which has a large number of particles N, and whose density slowly varying. Such a system necessarily occupies a large volume and has a dense spectrum in the sense that a characteristic energy spacing, Δ, between many body energies is very small compared to the total (or kinetic) energy of the ground state. In that case we have the following kind of situation.

(1.) For $M \leq 10$, say, $E^M \approx E_1$; $k\theta, k\theta_s << E_1$, and hence $E_{xc}^M \approx E_{xc}^1$. This is precisely what Eqs. (24)-(26) will yield, even though the assumption "M sufficiently large" is not satisfied.

(2.) For $M > 10$, thermodynamics (equivalence of different ensembles) becomes valid.

Thus we see that the explicit assumption of "M sufficiently large" is in fact not needed (the results hold down to $M = 1$) provided only that $n(r)$ is smooth enough and the number of particles is sufficiently large.

These are the same assumptions logically needed for the LDA for ground states. However experience has shown that the ground state LDA gives very useful results even for $N = 1$ and 2 and even if $n(r)$ is not very slowly varying. We may expect a similar experience for the present quasi-LDA for excited states. We hope to have results for some physically interesting systems (initially atoms) in the near future.

REFERENCES

[1.] W. Kohn, P. Vashishta: in Theory of the Inhomogeneous Electron Gas, ed. by S. Lundquist and N. March, (Plenum Press, New York, 1983).

[2.] P. Hohenberg and W. Kohn, Phys. Rev. **136**B, 864-871 (1964).

[3.] N. D. Mermin, Phys. Rev. **137**A, 1441-1443 (1965).

[4.] W. Kohn and L. U. Sham, Phys. Rev. A **140**, 1133-1138 (1965).

[5.] A. K. Theophilou, J. Phys. C, **12**, 5419-5430 (1979).

[6.] W. Kohn, Phys. Rev. B, to be published.

Attenuation of SAW Due to Electron Phonon Interaction

Moises Levy† and Susan C. Schneider

INTRODUCTION

Electron phonon attenuation produced by sound waves propagating in bulk materials has been studied extensively, both theoretically [1-5] and experimentally [6]. However, the computation of the electron phonon attenuation produced by surface acoustic waves (SAW) propagating in a thin metallic film deposited on a piezoelectric substrate is not as straightforward as for bulk waves, and no ab initio calculations similar to those of HOLSTEIN [1] and PIPPARD [2] exist for electron phonon interaction in thin films. There are, however, several models that incorporate the results of these earlier works to estimate SAW electron phonon attenuation in the limit where $ql \ll 1$, where q is the propagation vector of the sound wave and l is the electron mean free path [7-10]. Three of these models will be summarized here [7,9,10].

NORMAL STATE SAW ELECTRON PHONON ATTENUATION

The models to be summarized are those developed by TACHIKI, et al. [7], SALVO, et.al. [9] and SNIDER, et.al. [10]. These models all share some initial, rather stringent, assumptions used to calculate the electron phonon contribution to the attenuation of SAW propagating in a metallic film deposited on a piezoelectric substrate. These similar assumptions are (a) the attenuation is only produced in the metallic film, (b) the piezoelectric interactions between the film and substrate are negligible and (c) the film and substrate are both isotropic elastic bodies. The geometry used in all calculations is the same and is shown in Fig. 1. The SAW with wavevector, q, propagates along the x-axis, with the z-axis perpendicular to the film plane. The film/substrate interface is the z=0 plane and the free surface of the film is the plane z=h. In addition, all models are valid only in the $ql \ll 1$ limit. The models differ essentially only in how the SAW motion is decomposed into compressive and shear motion with the first being assigned longitudinal type attenuation and the second transverse; and in the ease with which the differing analytic approaches may be used to calculate the SAW electron phonon attenuation.

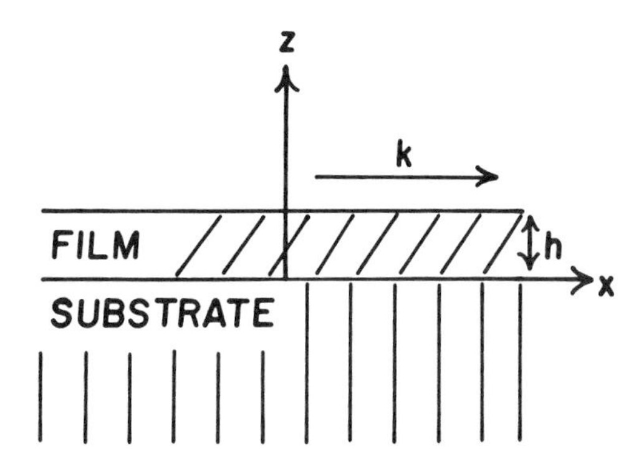

Fig. 1. Geometry used for surface acoustic wave attenuation calculation. Surface wave propagates in $\vec{x}$ direction.

Since all these models rely on the results of calculations of the electron phonon attenuation of bulk sound waves, the pertinent equations are included here for completeness. In the case of longitudinal waves, in the limits $\omega\tau \ll 1$ and $ql \ll 1$, the attenuation per unit length takes the form,

$$\alpha = 4Nmv_F^2\omega^2\tau/(15\rho v_l^3) \qquad (1)$$

where N is the number of electrons per unit volume, v_F is the Fermi velocity, ρ is the material density, τ is the electron relaxation time and v_l is the velocity of longitudinal sound waves, $\omega = qv_l$ is the angular frequency of the applied sound wave. For transvere sound waves in the limit where the wavelength of the sound wave is longer than the classical skin depth, the attenuation per unit length is given by (again in the limits where $\omega\tau \ll 1$ and $ql \ll 1$)

$$\alpha = Nmv_F^2\omega^2\tau/(5\rho v_t^3) \qquad (2)$$

In this equation, v_t is the velocity of the transverse sound wave, with $\omega = qv_t$.

In the model developed by TACHIKI, et al. [7] it is assumed that the presence of the metallic thin film does not affect the displacement fields of the SAW in the substrate and that the particle motion across the film/substrate interface is continuous. The expression obtained for the SAW attenuation is thus valid in the limit $qh \ll 1$, where h is the thickness of the film. By regarding the particle motion of the SAW in the x-direction as compressive and that in the z-direction as shear in nature, the compressive strain energy loss, which yields (1) for longitudinal bulk waves, is assigned to the x-component of the SAW wave while the shear energy loss, which yields (2) for transverse bulk waves, to the z-particle motion. The attenuation of the SAW is then obtained from the ratio of the sum of energy loss for compressive and shear strains divided by the energy of the surface wave in the substrate. The expression for the attenuation of SAW due to electron phonon interactions which was obtained is

$$\alpha = \frac{Nmv_F^2\omega^2\tau}{\rho v_s^3}(qh)F(\nu) \tag{3}$$

where v_s is the SAW velocity and $F(\nu)$ is a function of Poisson's ratio, ν, for the substrate. The results obtained from this equation agree closely with results from calculations which are valid for the whole range of qh [9,10]. A similar result for the attenuation was obtained when continuous strain across the film/substrate interface was considered to be the important boundary condition, differing only in the function for the dependence on Poisson's ratio of the substrate [8], however, the agreement between the continuous strain calculation with predictions from the models valid over the entire qh range is not as good.

As contrasted to the solution by TACHIKI, et al., the model developed by SALVO, et al. [9] takes into account both the motion of the substrate and the film. He followed the procedure of FARNELL and ADLER [11] to find the partial wave solution for both the substrate and the film. The value of the surface wave velocity and the six expansion coefficients of the partial waves are then determined by using the following boundary conditions for the x- and z-components; (i) the surface of the film is stress free, (ii) stress is continuous across the film/substrate interface and (iii) the amplitude of the SAW is continuous across the interface. The resultant six equations may be written in matrix form as,

$$\begin{pmatrix} a & 1 & -a_F & -1 & -a_F & -1 \\ -1 & -d & 1 & d_F & -1 & -d_F \\ \mu A/\mu_F & 2\mu d/\mu_F & -A_F & -2d_F & A_F & 2d_F \\ -2\mu a/\mu_F & -\mu A/\mu_F & 2a_F & A_F & 2a_F & A_F \\ 0 & 0 & A_F e^{a_F qh} & 2d_F e^{d_F qh} & -A_F e^{-a_F qh} & -2d_F e^{-d_F qh} \\ 0 & 0 & -2a_F e^{a_F qh} & -A_F e^{d_F qh} & -2a_F e^{-a_F qh} & -A_F e^{-d_F qh} \end{pmatrix} \begin{pmatrix} C_1 \\ C_2 \\ C_3 \\ C_4 \\ C_5 \\ C_6 \end{pmatrix} = 0, \tag{4}$$

where the subscript F refers to film parameters and those parameters without subscripts are evaluated for the substrate. For the substrate, $a = (1 - v^2/v_t^2)^{1/2}$, and $d = (1 - v^2/v_l^2)^{1/2}$, where $v_l = [(\lambda + 2\mu)/\rho^2]^{1/2}$ and $v_t = [\mu/\rho^2]^{1/2}$ are the bulk longitudinal and transverse velocities of the substrate, v is the SAW velocity and $A = (1 + a^2)$. Similar definitions are used for a_F, d_F and A_F, where the velocities v_{lF} and v_{tF} are the bulk velocities of the film material. The only undetermined parameter in this matrix is the SAW velocity, v. A computer is used to find that value of the velocity for a particular film/substrate combination which makes the determinant of this matrix vanish (non-trivial solution). Once the velocity of the SAW is known, the partial wave coefficients may be found. The procedure then followed to calculate the electron phonon attenuation is similar to the approach used by TACHIKI, et. al., described above, except that actual integrations are performed for the film as well as for the substrate. The attenuation per unit length obtained with this calculation can thus be extended to any value of qh provided that the limit of $ql \ll 1$ is still observed.

The two calculations mentioned above solved for the displacement field of the lattice by neglecting losses and then evaluated the attenuation by assigning compressive strain energy losses to the x-component and shear energy losses to the z-component of the particle motion of the SAW wave. SNIDER, et al. [10] avoid this type of approximation by utilizing a macroscopic equation of motion proposed by MASON [3] which includes viscous loss terms and is valid for $ql \ll 1$. Assuming a harmonic time dependence for the particle motion, $\vec{\Psi}(\vec{r},t) = \vec{\Psi}(\vec{r})\, e^{-i\omega t}$, the equation of motion including viscous loss may be written as

$$-\rho\omega^2\vec{\Psi} = [\lambda + \mu - i\omega(\chi + \eta)]\, \vec{\nabla} \cdot (\vec{\nabla} \cdot \vec{\Psi}) + (\mu - i\omega\eta)\, \vec{\nabla}^2\vec{\Psi} \tag{5}$$

where λ and μ are the isotropic Lame constants, χ and η are the corresponding viscosity coefficients and ρ is the density of the material. This equation is the same as that of a non lossive medium where λ is replaced by $(\lambda - i\omega\chi)$ and μ by $(\mu - i\omega\eta)$. Hence, the solutions for the displacement fields and the propagation velocity are the analytic continuation of the solutions to the equation of motion without explicit losses. Expressions for χ and η were developed by MASON [5] by considering the metal as a free electron gas. The bulk coefficient of viscosity, K, is related to the compressional viscosity, χ, and shear viscosity, η, through the equation $K = \chi + 2/3\,\eta$. Since for a monatomic gas, such as a free electron gas, the bulk coefficient of viscosity is equal to zero because uniform compressions in such a system produce no viscous losses, $\chi = -2\eta/3$. Furthermore, $\eta = Nmv_F^2\tau/5$, as derived by MASON [5].

Again using the procedure of FARNELL and ADLER [11], with (5) used to describe particle motion in the metallic film, a matrix identical in form to that of (4) results, this time, however, with complex parameters a_F and d_F since for the film $v_{lF}=[(\lambda+2\mu-i4\omega\eta/3)/\rho^2]^{1/2}$ and $v_{tF}=[(\mu-i\omega\eta)/\rho^2]^{1/2}$ are now the longitudinal and transverse bulk velocities. Thus the viscous terms make the coefficients in the matrix complex. Again the SAW velocity is the only undetermined parameter in this matrix. Solving for that value of v which makes the determinant vanish results in a complex value for velocity and the attenuation per unit length can be calculated from this velocity as the imaginary part of (ω/v). Thus in the procedure used by SNIDER, et al. [10], it is not necessary to solve explicitly for the partial wave coefficients in order to calculate the attenuation. In addition, the fact that displacements along the x- and z-axes contain both compressive and shear components has been taken into account in the equation of motion.

The attenuation coefficient of SAW as a function of qh for an Aluminum film on either a quartz or a $LiNbO_3$ substrate as calculated by SNIDER, et al. is shown in Fig. 2. The value of η for Al used for this calculation was 2.19×10^{-3} poise. The open squares are for Al on a $LiNbO_3$ substrate, the filled squares are for an Al film on a quartz substrate. Also shown in this figure as filled circles are the results obtained for an Al film on a $LiNbO_3$ substrate using the method developed by SALVO, et al. [9]. It may be seen in this figure, that the approximation which assigns compressive strains to the x-motion and shear strains to the z-motion and does not have to solve a complex matrix gives results similar to those obtained from the calculation of SNIDER, et al, which does require that a complex matrix be evaluated. In the limit of $qh \ll 1$, the attenuation is a linear function of qh for both calculations, as was the case for the calculations by TACHIKI, et al. [7].

SUPERCONDUCTING STATE SAW ATTENUATION

In 1957, BARDEEN, COOPER and SCHRIEFFER (BCS) [12] derived a famous expression for the ratio of the ultrasonic attenuation coefficient in the superconducting state, α_s, to that in the normal state, α_n, which was valid for longitudinal waves in the $ql \gg 1$ limit. Their result was

$$\frac{\alpha_s}{\alpha_n} = \frac{2}{e^{\Delta/kT} + 1} \tag{6}$$

where 2Δ is the superconducting energy gap, k is Boltzmann's constant and

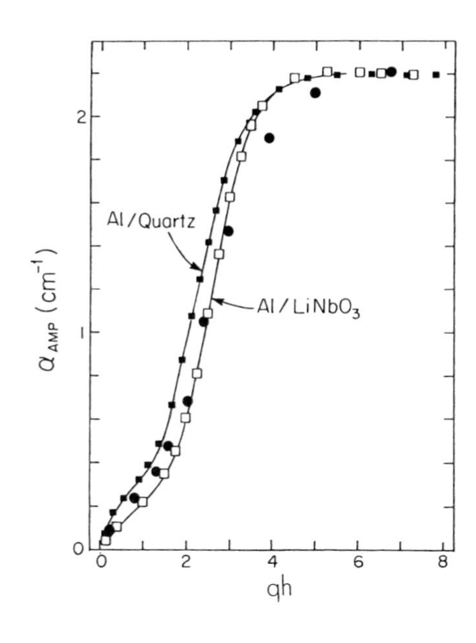

Fig. 2. Amplitude attenuation of SAW as a function of qh for Aluminum films on quartz (■ [10]) and $LiNbO_3$ (□ [10] and ● [9]).

T is the absolute temperature. Later it was shown by LEVY [5] using a Boltzmann transport method similar to that used by HOLSTEIN [1] for the attenuation of sound waves in the normal state, that the BCS result is also valid for both longitudinal and transverse waves in the limit of $ql \ll 1$. It was shown in the previous section that the attenuation of SAW due to electron phonon interactions in the normal state of a metallic film may be attributed to the shear viscosity of the electron gas. Since this shear viscosity of the electron gas follows the BCS relation as derived for transverse sound waves which are purely shear in nature, it should follow therefore that the attenuation of a surface acoustic wave in the $ql \ll 1$ limit should also follow the BCS relation where α_s and α_n are now the superconducting and normal state SAW electron phonon attenuation. The same result is obtained from models which ascribe the attenuation of the SAW to the sum of the compressive and shear losses, since both of these losses also follow the BCS relation in the limit $ql \ll 1$. Experimental confirmation of (6) for SAW can be seen in the work of AKAO [13] on films of Pb and In, ROBINSON, et al. [14] on Al films, BAILEY, et al. [15] for Zn films and FREDRICKSEN, et al. [16] for Nb_3Sn films.

CONCLUSION

At present, while there are no ab initio calculations of the electron phonon contribution of the attenuation of SAW, several models exist which can be used to estimate this attenuation. While the models all rely on concepts used to calculate electron phonon attenuation for bulk sound waves in the limit where $ql \ll 1$, they differ in how the SAW particle motion is interpreted as compressive and shear for the purposes of assigning bulk attenuation behavior and in the ease with which they may be used to calculate the SAW electron phonon attenuation. In the limit $qh \ll 1$, all three models predict that the attenuation is a linear function of qh, and when there is not a significant difference in the material parameters of the film and the substrate, all three models yield similar values for the attenuation. It is only when the acoustic impedances of the film and the substrate are significantly different that any meaningful differences in the value of the attenuation predicted by the three models appear. It would be desirable to have a theoretical model for the attenuation of SAW in the $ql \gg 1$ limit, for both the normal and superconducting states. In this limit, the bulk modulus of viscosity cannot be set equal to zero since the attenuation coefficients for longitudinal and transverse waves are no longer related by a factor of 4/3 times the appropriate ratio of the corresponding sound velocities as they are expected to be in the limit $ql \ll 1$. It is possible to obtain $ql \gg 1$ for SAW propagating on the surface of very pure single crystal metals. This would require launching the SAW directly on the metal either by depositing a piezoelectric transducer on the surface of the metal or by wedge bonding a transducer to the metal to excite the surface wave.

REFERENCES

† Research supported by Air Force Office of Scientific Research under AFOSR Grant no. 84-0350.

1. T. Holstein: Westinghouse Research Memo, 60-94698-3-M17 (1956), unpublished.
2. A.B. Pippard: Phil. Mag. 46, 1104 (1955).
3. W.P. Mason: Phys. Rev. 97, 557 (1955).
4. R.W. Morse: Phys. Rev. 97, 1716 (1955).
5. Moises Levy: Phys. Rev. 131, 1497 (1963).
6. See for example, H.E. Bömmel: Phys. Rev. 96, 220 (1954). R.W. Morse and H.V. Bohm: Phys. Rev. 108, 1094 (1957). J.L. Brewster, M. Levy and I. Rudnick: Phys. Rev. 132, 1062 (1963). M. Levy and I. Rudnick: Phys. Rev. 132, 1073 (1963).
7. M. Tachiki, H. Salvo, Jr., D.A. Robinson and M. Levy: Sol. St. Commun. 17, 653 (1975).
8. Moises Levy, Harry Salvo, Jr., David A. Robinson, Kazumi Maki and Masashi Tachiki: Proc. 1976 Ultrasonics Sym. (ed. by J. DeKlerk) (IEEE, New York, 1976), p. 633.
9. H.L. Salvo, Jr., M. Levy: Proc. 1978 Ultrasonics Symp. (ed. by J. deKlerk) (IEEE, New York, 1978), p. 400.
10. Dale R. Snider, Hans P. Fredricksen and Susan C. Schneider: J. Appl. Phys. 52, 3215 (1981).

11. G.W. Farnell and E.L. Adler: In Physical Acoustics, Vol. IX, ed. by W.P. Mason and R.W. Morse (Academic Press, New York, 1972), p. 35.
12. J. Bardeen, L.N. Cooper, and J.R. Schrieffer: Phys. Rev. 108, 1175 (1957).
13. F. Akao: Phys. Lett. 30A, 409 (1969).
14. D.A. Robinson, K. Maki and M. Levy: Phys. Rev. Lett. 32, 13 (1974).
15. W. Bailey and B. Marshall: Phys. Rev. B19, 3467 (1969).
16. H.R. Fredricksen, H.L. Salvo, Jr., M. Levy, R.H. Hammond, and T.H. Geballe: Phys. Lett. 75A, 389 (1980).

Close-Coupling Calculations of H_2/Cu (001) Scattering

B. H. Choi, N. L. Liu, and X. Shen

Abstract

Simultaneous vibrational and rotational close-coupling calculations are performed to obtain the rotational excitation probabilities in H_2/Cu (001) rigid surface scattering. A semiempirical LEPS potential is used for H_2/Cu interaction. The rotational excitation probabilities are compared with the corresponding rigid rotor close-coupling calculations. Structures of the Feshbach type resonance appear in the energy dependence of both the elastic scattering and rotational excitation probabilities in the non-rigid rotor calculations, whereas no such structure is found in the rigid rotor case. This result is explained by the temporary adsorption of the molecule on the surface due to the lowering in the molecule-surface potential as the molecule is stretched. This calculation demonstrates the inadequacy of the rigid rotor approximation in describing the phenomenon of selective adsorption.

I. Introduction

The dynamics of molecule-surface scattering has often been treated by quasiclassical trajectory (QCT) calculations or sudden approximation. Such methods have rather limited degree of accuracy and range of validity. The exact quantum mechanical close-coupling approach was first applied to rotationally inelastic molecule-surface scattering for a rigid flat surface,[1] and extended recently to a rigid corrugated surface by Drolshagen et. al.[2] By computing the rotation-diffraction transition probabilities, these authors concluded that the coupling between rotations and diffractions is insignificant so that both can be treated independently to a good approximation.

In all the prior close-coupling calculations, the incident molecule has been treated as a rigid rotor. In this paper, we carry out a close-coupling calculation which includes both the molecular rotational and vibrational degrees of freedom for H_2 scattering from a Cu (001) surface. For simplicity, we take the surface to be rigid. Since the diffraction and rotation transitions were found to be essentially independent,[2] we further simplify our calculation by assuming the surface to be flat, uncorrugated. We compare here the rotational transition probabilities obtained from the rigid rotor approximation to those from non-rigid rotor approximation where rotational and vibrational states are simultaneously coupled. In this manner, the importance of vibrational effect (which corresponds to closed channels here) can be determined. We present in this paper only calculations of rotationally elastic and inelastic scattering cross sections. Close-coupling calculation applied to extensive selective adsorption study in the same system is reported elsewhere[3] together with our QCT study[4] and quantum mechanical calculation of the quasibound states[5] for selective adsorption.

II. Method of Computation

The Hamiltonian of the diatomic molecule-flat, rigid (phononless) surface system is given by

$$H = \frac{-\hbar^2}{2M} \nabla_r^2 - \frac{\hbar^2}{2\mu} \nabla_\rho^2 + V_M(\rho) + V_{M-S}(z,\vec{\rho}) \tag{1}$$

where M is the mass of the molecule, μ is the reduced mass between atoms in the molecule, $\vec{r} = z\,\hat{z} + \vec{R}$ is the position vector of the center of mass of the molecule, $\vec{\rho}$ is the relative vector between the two atoms in the projectile, and $\hat{z}$ is the outward unit vector normal to the surface, $\vec{R}$ is parallel to the surface, V_M is the intra-molecular potential, and V_{M-S} is the molecule-surface interaction potential. The Schrödinger equation for the molecule-surface system is

$$H\Psi_i = E\Psi_i \tag{2}$$

with the asymptotic boundary condition

$$\Psi_i \underset{z\to\infty}{\sim} \exp(i\vec{k}_i.\vec{r})\ \frac{1}{\rho}\ \phi_{v_i j_i}(\rho)\ Y_{j_i m_i}(\hat{\rho}) + (\text{scattered wave}) \tag{3}$$

where $\vec{k}_i$ is the incident momentum of the projectile and $\frac{\phi_{vj}}{\rho} Y_{jm}$ is the vibrational and rotational wave function of the molecule with vjm denoting vibrational and rotational quantum numbers and the rotational projection. The subscript i denotes the initial condition and E is the total energy given by

$$E = \frac{\hbar^2 k_i^{\,2}}{2M} + \varepsilon_{v_i j_i} \tag{4}$$

with ε_{vj} the vibrational and rotational energy of the projectile. It is also convenient to decompose $\vec{k}_i$ into components perpendicular and parallel to the surface,

$$\vec{k}_i = k_{v_i j_i}\,\hat{z} + \vec{K}_i \qquad\qquad (k_{v_i j_i} = k_{iz}) \tag{5}$$

The dependence of the potential on the azimuthal angle of $\vec{\rho}$ is negligible and thus we can put

$$V_{M-S}(z,\vec{\rho}) = \sum_{\lambda} V^{\lambda}_{M-S}(z,\rho)\ Y_{\lambda o}(\hat{\rho}) \tag{6}$$

This approximation implies the azimuthal quantum number m of the molecule is conserved. Because of the flatness of the surface and the m-conservation, it is sufficient to consider only the $\vec{G}$=0 and m=0 components of the wave function. Here, $\vec{G}$ is the two dimensional reciprocal lattice vector. Therefore, Ψ_i is expanded as

$$\Psi_i = e^{i\vec{K}_i.\vec{R}} \sum_{vj} \Psi_{vj,\,v_i j_i}(z)\ \frac{\phi_{vj}(\rho)}{\rho}\ Y_{jo}(\hat{\rho}) \tag{7}$$

Substituting Eq. (7) into Eq. (2) and projecting onto the basis states, we have

$$\left(\frac{d^2}{dz^2} + k_{vj}^2\right) \Psi_{vj,\ v'j'}(z) \tag{8}$$

$$= \frac{2M}{\hbar^2} \sum_{\lambda v''j''} V^{\lambda}_{vj,\ v''j''}(z)\ \Psi_{v''j'',\ v'j'}(z)$$

Here, $V^{\lambda}_{vj,\ v'j'}(z) = \left[\frac{(2j+1)\ (2j'+1)}{4\pi\ (2\lambda+1)}\right]^{1/2} \quad |\langle joj'0|\lambda o\rangle|^2$

$$\times \int d\rho\ \phi_v(\rho)\ V^{\lambda}_{M-S}(z,\rho)\ \phi_v(\rho) \tag{9}$$

$$= V^{\lambda}_{v'j',\ vj}(z)$$

$$\text{and } k_{vj}^2 = \frac{2M}{\hbar^2}(E-\varepsilon_{vj}) - K_i^2 \tag{10}$$

We put $k_{vj} > 0$ when $k_{vj}^2 > 0$ (open channel), and $k_{vj} = i\kappa_{vj}$ when $k_{vj}^2 < 0$ (closed channel). It is seen that $\vec{K}_i$ is a constant motion. We consider here only normal incidence and thus $\vec{K}_i = 0$. The boundary condition of Eq. (3) results in

$$\Psi_{vj,\ v'j'}(z) \underset{z\to\infty}{\sim} \left[\frac{k_{v'j'}}{k_{vj}}\right]^{1/2} [\delta_{vv'}\delta_{jj'}\exp(-ik_{vj}z) - S_{vj,v'j'}\exp(ik_{vj}z)] \quad \text{(open)} \tag{11}$$

$$\underset{z\to\infty}{\sim} -\bar{S}_{vj,\ v'j'}\exp(-\kappa_{vj}z) \quad \text{(closed)}$$

This is called the scattering matrix boundary condition.

In practice, we employ the reactance matrix boundary condition in order to make the numerical scattering matrix $S = (S_{vj,\ v'j'})$ unitary. The probability of the vibrational and rotational transition is given by

$$P_{v'j'\to vj} = |S_{vj,\ v'j'}|^2$$

III. Results and Discussions

Integration of the coupled differential equation given by Eq. (8) was carried out from z = 1.0 to $17a_o$. Numerical procedures were adopted from Ref. 6. For H_2 - Cu surface interaction potential, the semiempirical LEPS potential in Ref. 7 was employed. Two reasons motivated us to use this potential. First, we have made extensive QCT calculations based on this potential.[4] Thus, we could compare the present quantal results to those from classical approach. Secondly, the vibrational degrees of freedom can be easily incorporated in this potential.

Because of the m-conservation, the number of coupled channels is considerably reduced. It is simply given by the number of vibrational states multiplied by the number of rotational levels. Furthermore, even and odd rotational states do not mix with each other. Therefore, our close-coupling calculations for even and odd rotational states are performed separately.

As previously mentioned, we carried out the computations for the rotational transition probabilities of the H_2-Cu scattering with rigid rotor approximation and without this approximation. In the latter case, simultaneous vibrational and rotational close-coupling calculations were made. Computations were carried out in the region of incident energies E = 0.1 ~ 0.3 eV.

Figure 1 shows the elastic scattering probabilities, (v,j) = (0,0) → (0,0) transition, with the rigid rotor approximation. These result from the close-coupling calculations with inclusion of v = 0, j = 0, 2, 4, 6 states. It is well known that close-coupling calculations within the ground vibrational manifold is basically equivalent to the rigid rotor approximation. The elastic scattering probabilities monotonically decrease as E increases. This energy dependence was also obtained in other rigid rotor calculation for H_2-LiF surface system.[2]

The elastic scattering probabilities obtained from the non-rigid rotor approximation are shown in Fig. 2. They tend to decrease as the incident energy E increases. We find resonance structures at around 0.103 and 0.187 eV, whereas no such structures appear in the rigid rotor case. The shapes of the resonances are such that at E = 0.103 eV, the non-

FIGURE 1

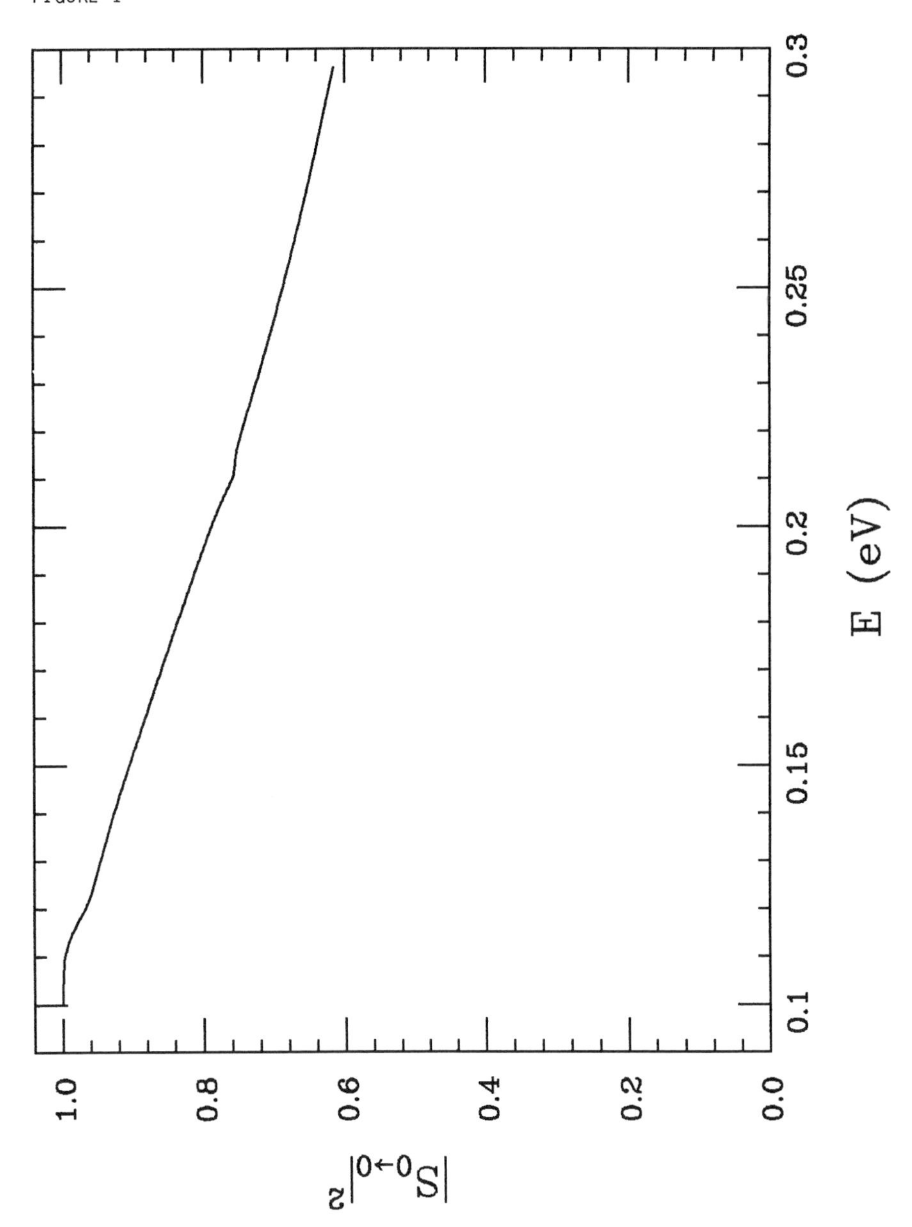

FIGURE 2

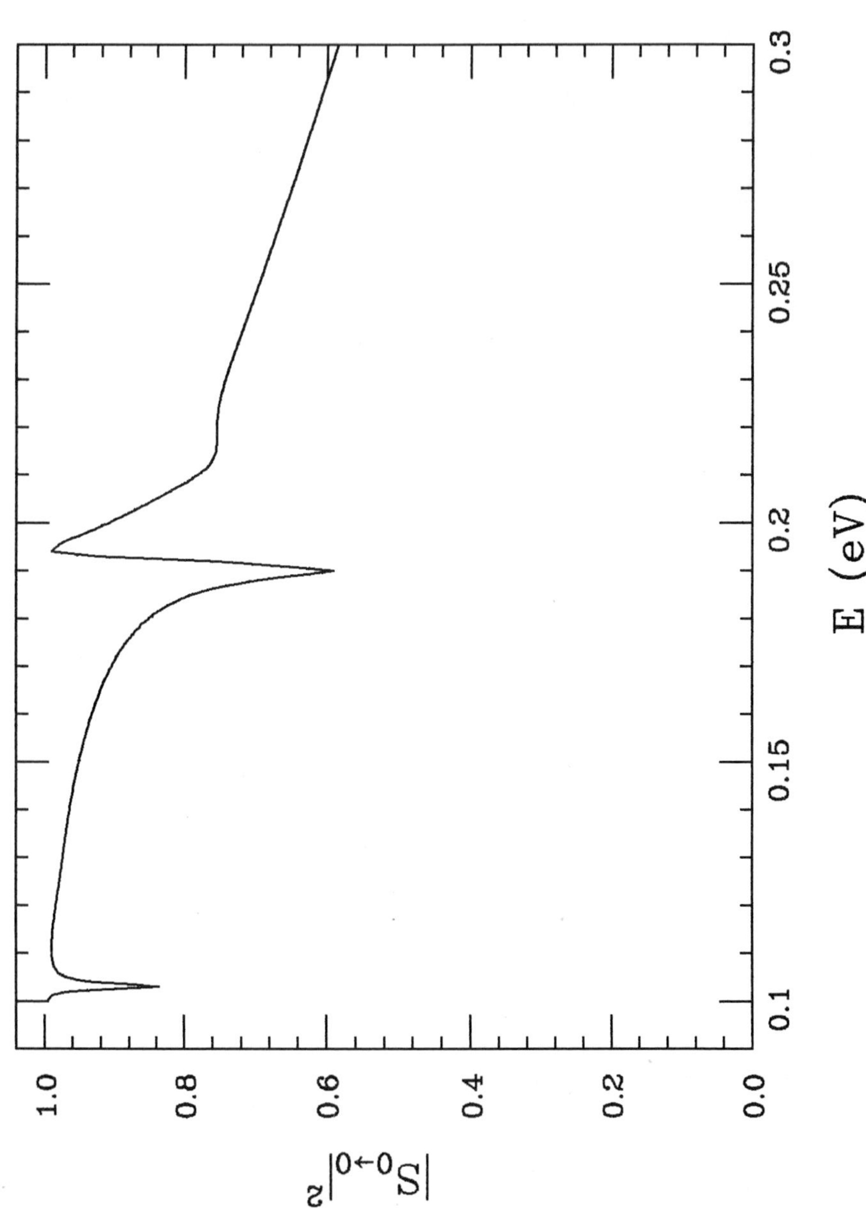

resonant background phase shift is $\pi/2$ radian and at E = 0.187 eV, it is $3\pi/4$ radian. These resonance structures are due to the fact that the molecule is temporarily adsorbed to the surface and subsequently desorbed in the scattering process. The reason this occurs in the non-rigid rotor scattering case is that the molecule can be adsorbed to the surface only by stretching the internuclear separation and thus decreasing the molecule-surface interaction potential. The rigid rotor molecule cannot be temporarily adsorbed to the surface in the present potential. Therefore, our calculation shows that the rigid rotor approximation is inadequate for describing the incident energy dependence of intensities/probabilities in the Feshbach type resonance scattering.

Figure 3 shows the $0\to2$ rotational excitation probabilities of the rigid rotor approximation resulting from the inclusion of the same states as in Fig. 1. They increase monotonically as E increases. Again, similar behavior was obtained in the calculation of H_2-LiF surface scattering with rigid rotor.[2] The $0\to2$ rotational excitation probabilities for the non-rigid rotor approximation are shown in Fig. 4. They result from the inclusion of the same states as in Fig. 2 in the close-coupling calculations. We again find the resonance structures at around 0.103 and 0.187 eV. Due to the normalization of the scattering matrix S originated from the unitarity condition, the resonance peaks of $0\to2$ excitation probabilities in Fig. 4 correspond to their respective dips in the $0\to0$ elastic scattering probabilities in Figure 2. These resonances in the $0\to2$ excitation can be explained in the same manner as the $0\to0$ elastic scattering. By stretching internuclear separation in the molecule, i.e., by the inclusion of higher vibrational states which belong to the closed channels in the present energy region, the molecule-surface potential energy is lowered. As a result, the molecule is temporarily bound to the surface and desorbed again at the resonance energy. It should be pointed out that such process cannot occur if the rigid rotor approximation is used.

We have also carried out the close-coupling calculations for odd rotational states with rigid and non-rigid rotor approximations. The v = 0, j = 1, 3, 5, 7 states were coupled in the rigid rotor approximation and v = 0, 1, 2, j = 1, 3, 5, 7 states are coupled in the non-rigid rotor approximation. As mentioned before, results from the inclusion of v = 0

FIGURE 3

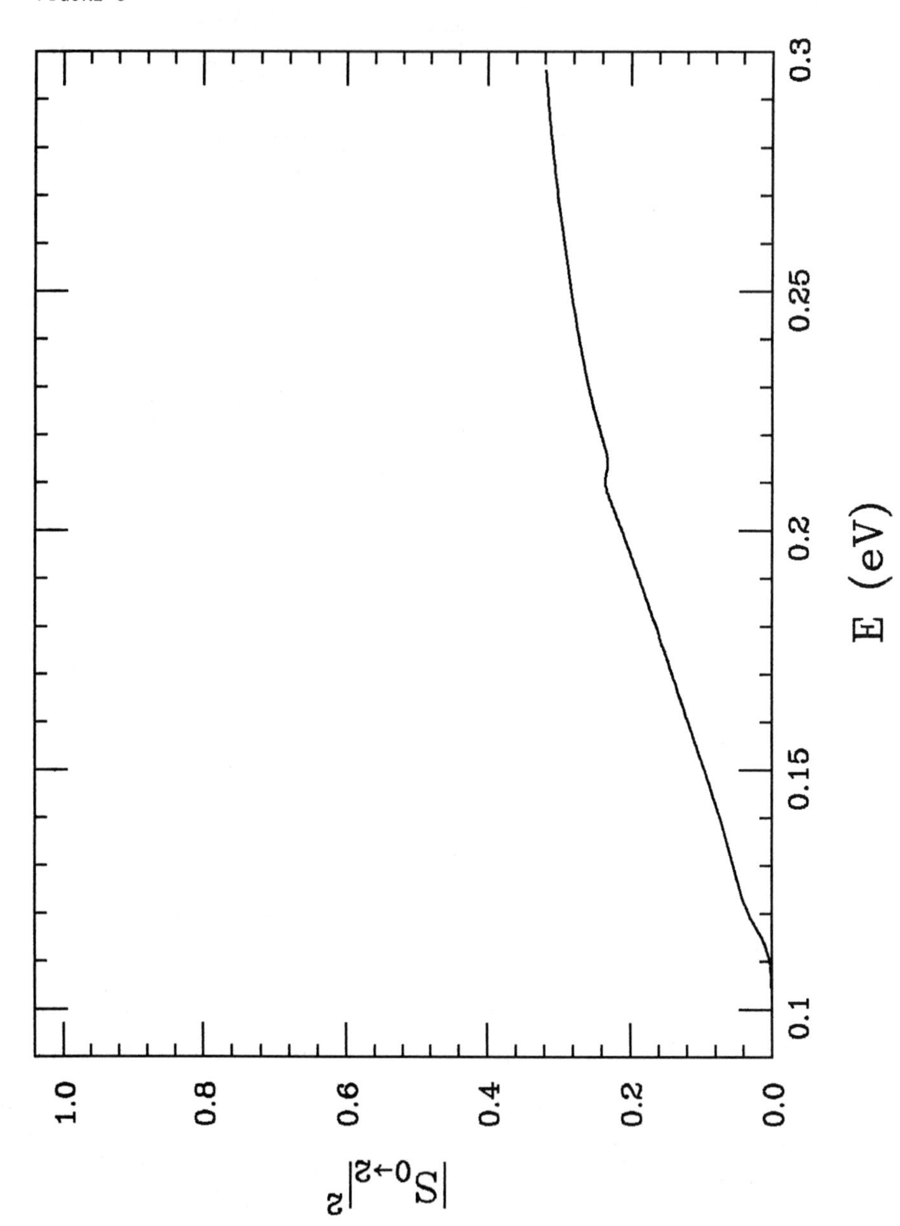

FIGURE 4

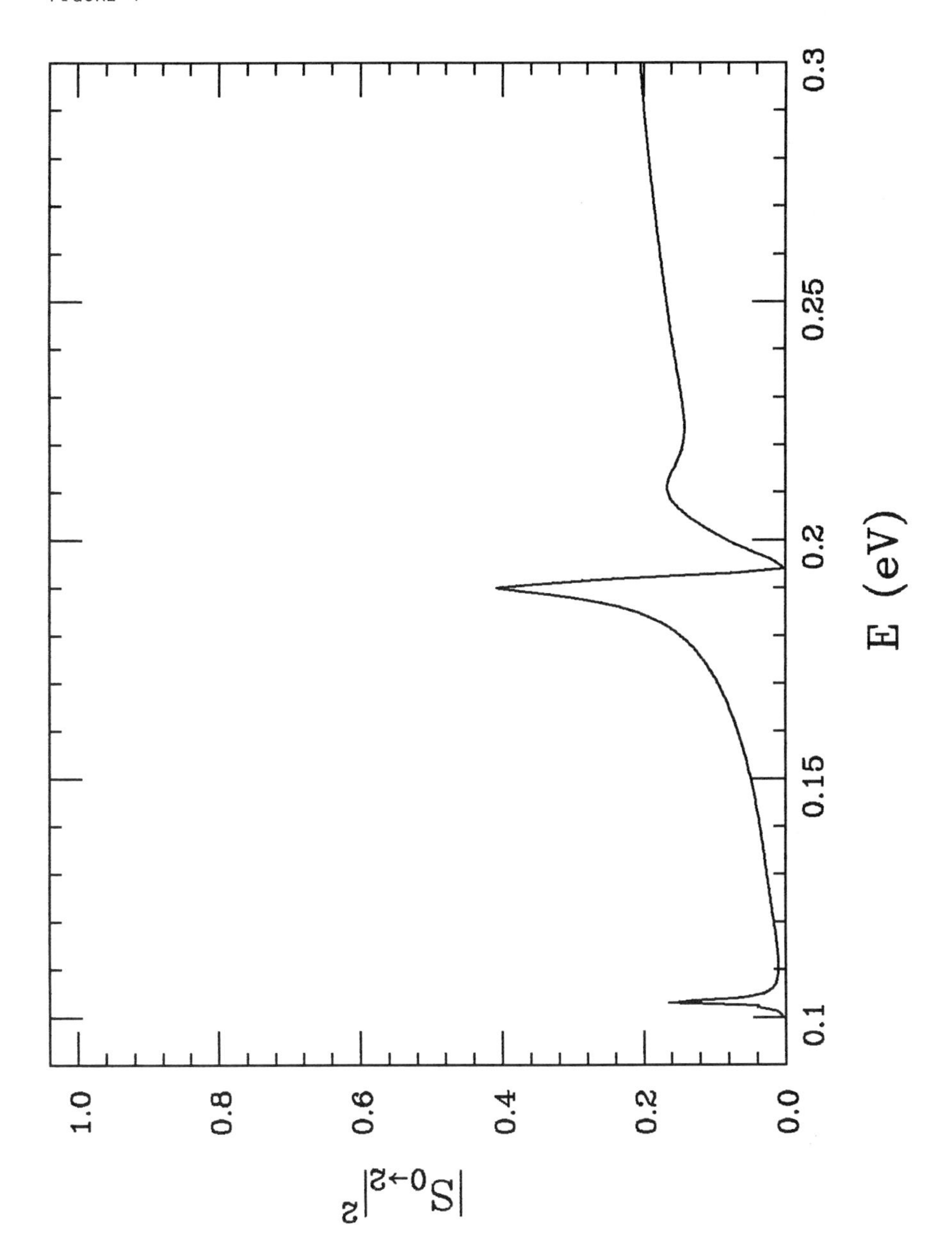

manifold only are very much close to those from the rigid rotor approximation. The 1→1 rotationally elastic scattering probabilities with rigid and nonrigid rotor approximations are shown in Figs. 5 and 6, respectively. In Figs. 7 and 8, the 1→3 rotational excitation probabilities are shown with rigid and non-rigid rotor approximations, respectively. The same remarks for the corresponding cases of the even rotational states are applicable to Figs. 5-8.

In summary, we have carried out the simultaneous vibrational and rotational close-coupling calculations for the elastic scattering and rotational excitation probabilities of H_2-Cu (001) collisions. This is the first such calculations in molecule-rigid surface scattering study. However, it is felt that complete convergence of the projectile's vibrational and rotational states may not have been achieved. Inclusion of much higher vibrational states would result in more substructures in the energy dependence of the probabilities, although major resonances around 0.187 eV in 0→0 transition most likely would not change substantially. Inclusion of much higher vibrational states will yield more deviation in the case of the rigid rotor calculations.

Our calculation clearly shows that in order to obtain the Feshbach type resonances in the rotational excitation/elastic scattering, inclusion of the vibrational states, which are closed channels in the low energy region, are crucial in the close-coupling calculations. The Feshbach resonances cannot be obtained from the rigid rotor approximation. Therefore, for molecule-surface scattering, rigid rotor approximation is not suitable for describing the selective adsorption phenomena which arise from the internal excitation of the projectile.

For comparison with experimental measurements, the use of more realistic potential and the inclusion of large number of vibrational states may be necessary in the close-coupling calculations. Computations in such direction are in progress.

Acknowledgment

This research project is supported in part by Universitywide Energy Research Group, University of California.

FIGURE 5

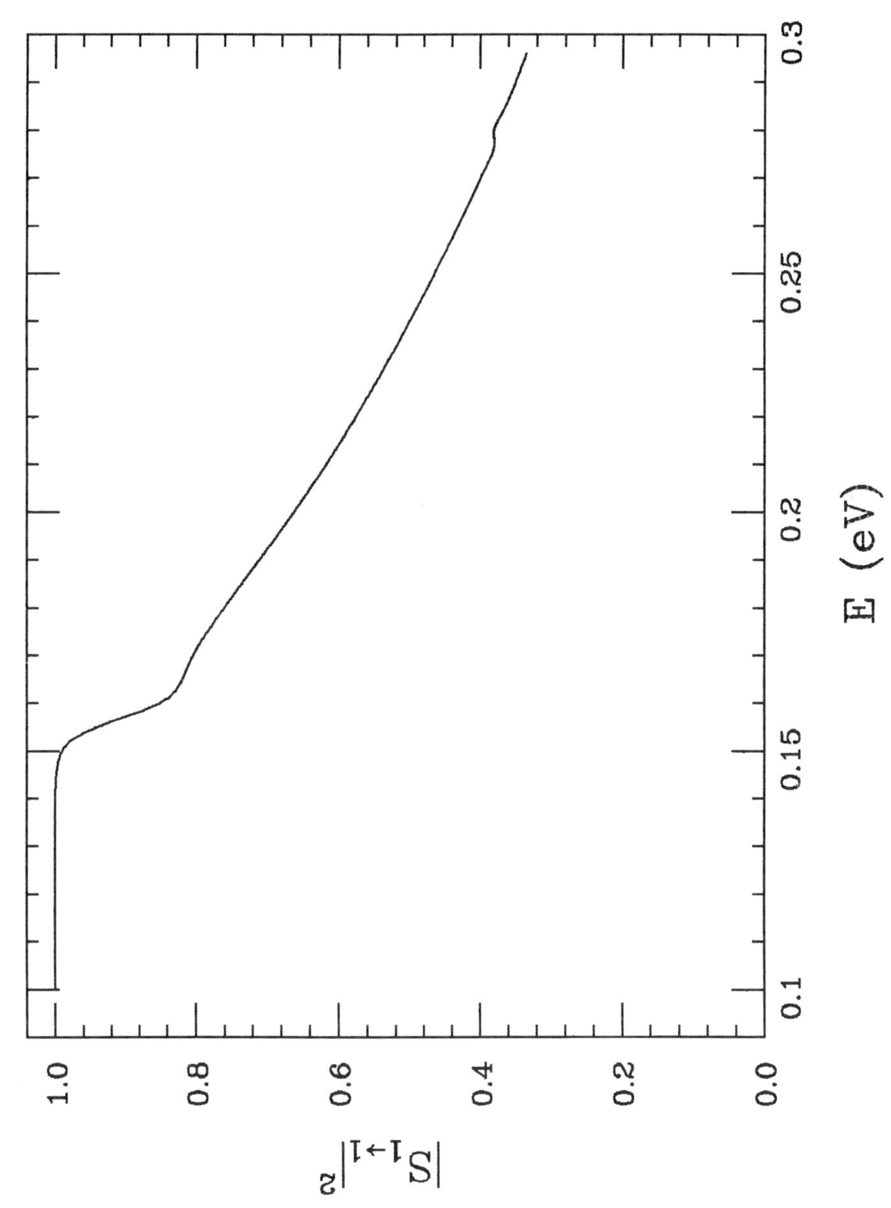

FIGURE 6

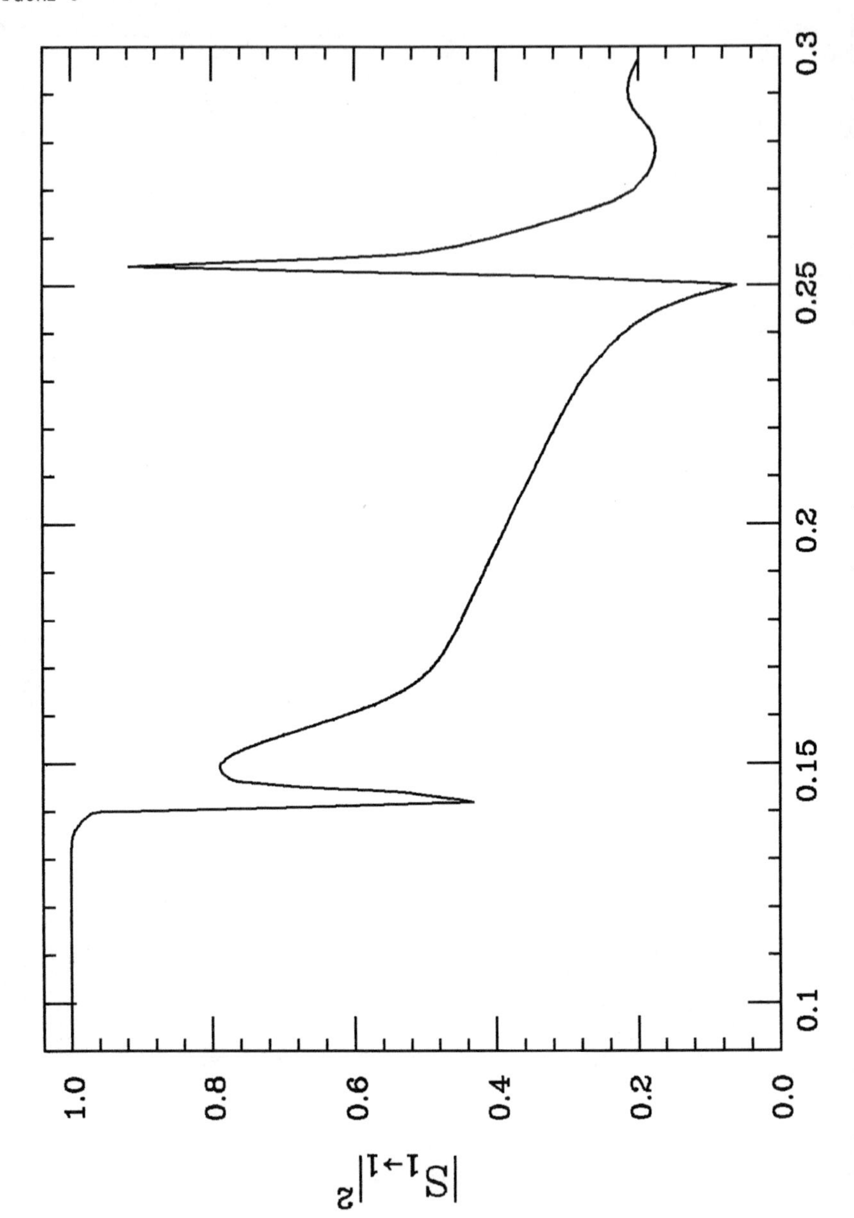

FIGURE 7

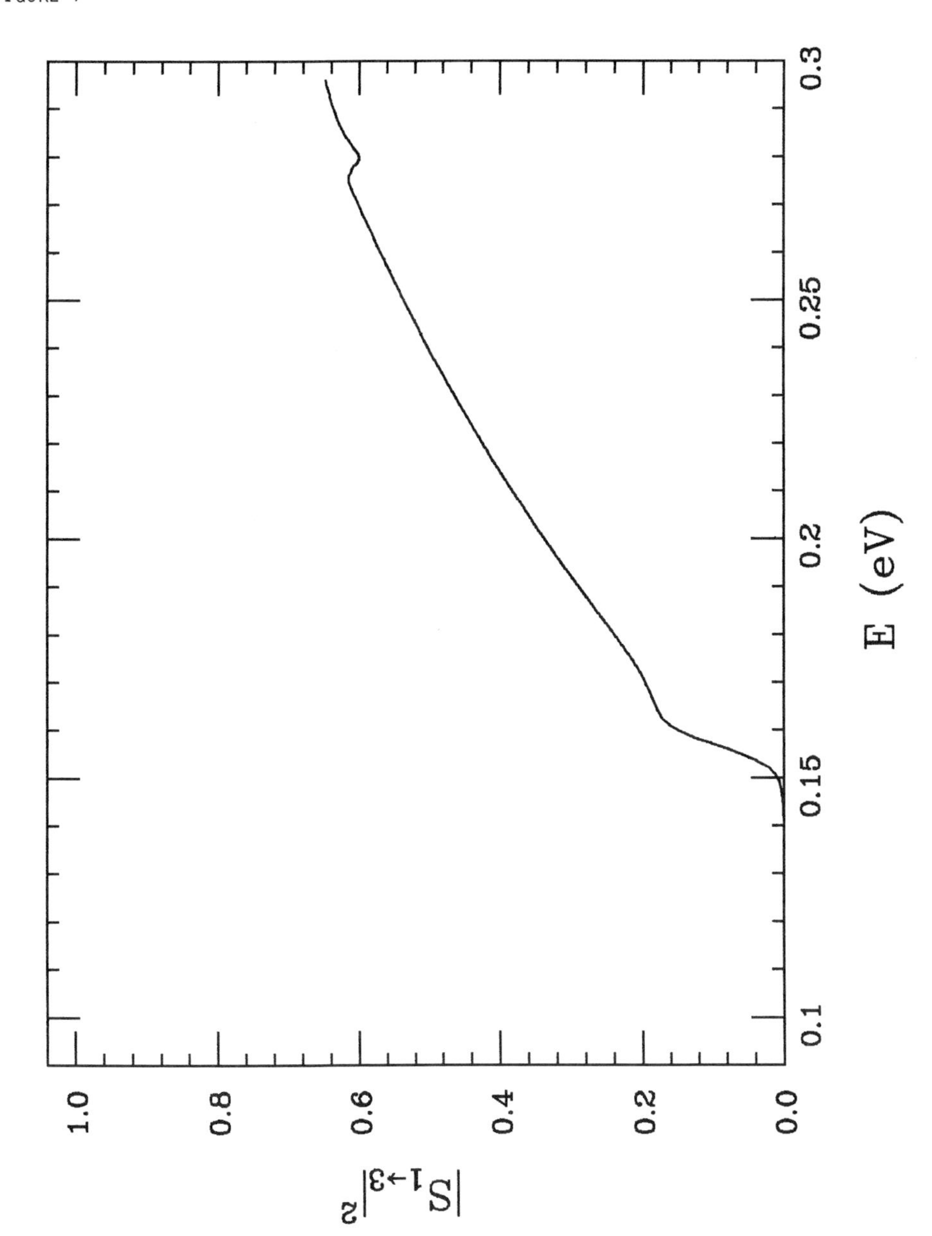

FIGURE 8

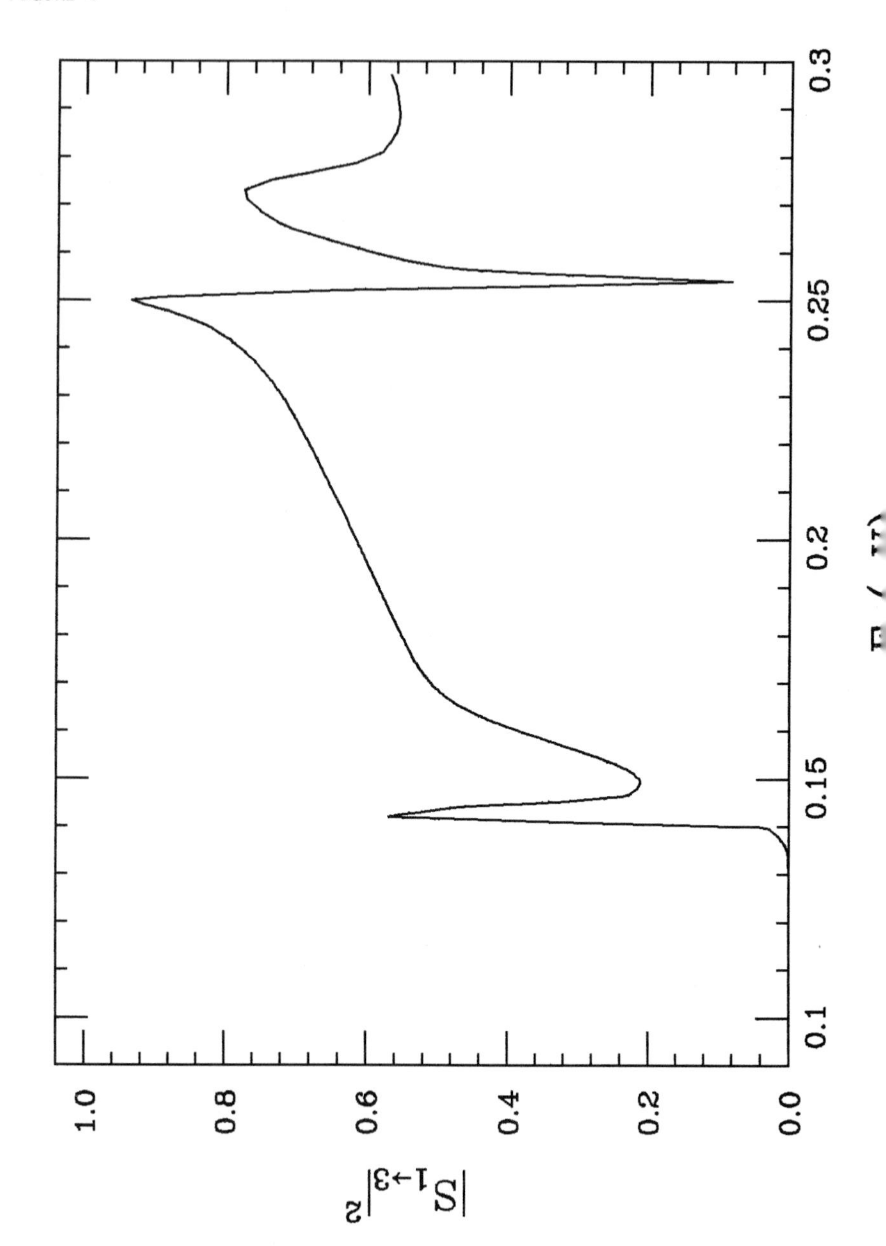

References

1. D.E. Fitz, L.H. Beard, and D.J. Kouri, Chem. Phys. 59, 257 (1981); R. Schinke, Chem. Phys. Lett. 87, 438 (1982); K. B. Whaley, J. C. Light, J. P. Cowin, and S. J. Sibener, Chem. Phys. Lett. 89, 89 (1982); J. A. Barker, A.W. Kleyn, and D.J. Auerbach, Chem. Phys. Lett. 97, 9 (1983).
2. G. DROLSHAGEN, A. Kaufhold, and J. P. Toennies, J. Chem. Phys. 83, 827 (1985).
3. B.H. Choi, N.L. Liu, and X. Shen, to be published.
4. N.L. Liu, B.H. Choi, and X. Shen, to be published.
5. B.H. Choi, N.L. Liu, and X. Shen, to be published.
6. B.H. Choi and K.T. Tang, J. Chem. Phys. 63, 1783 (1975).
7. A. Gelb and M.J. Cardillo, Surf. Sci. 75, 199 (1978).

Figure Captions

Figure 1. The (v,j) = (0,0) → (0,0) elastic scattering probabilities as a function of the incident energy E in the rigid rotor approximation.

Figure 2. Same as Fig. 1 in the non-rigid rotor approximation.

Figure 3. The (v,j) = (0,0) → (0,2) rotational excitation probabilities as a function of the incident energy in the rigid rotor approximation.

Figure 4. Same as Fig. 3 in the non-rigid rotor approximation.

Figure 5. The (v,j) = (0,1) → (0,1) elastic scattering probabilities as a function of the incident energy E in the rigid rotor approximation.

Figure 6. Same as Fig. 5 in the non-rigid rotor approximation.

Figure 7. The (v,j) = (0,1) → (0,3) rotational excitation probabilities as a function of the incident energy E in the rigid rotor approximation.

Figure 8. Same as Fig. 7 in the non-rigid rotor approximation.

Multiple Scattering in Semiconductor Quantum-Wells

S. K. Lyo

1. Introduction

The transport properties of a two-dimensional electron gas (2DEG) as found in semiconductor heterostructures, quantum-wells, and inversion layers are of much current interest. In 2DEGs the Fermi energy is small, typically a few tens of [meV] or less. It is well known from Holstein's transport theory of metals [1] that the Boltzmann equation with the single-scattering approximation is valid in the limit where the damping (Γ) of the electronic states near the Fermi surface is much smaller than the Fermi energy (ε_F). While the ratio Γ/ε_F is negligibly small in 3D metals, it is not necessarily so in a 2DEG. In this paper we analyze and assess the importance of multiple scattering corrections to the mobility of a 2DEG using Holstein's formalism [1]. We find that the two-scattering correction to the transport relaxation rate (TRR) for scattering by ionized impurities and acoustic phonons is of the order $(\Gamma/\varepsilon_F)\ell n(4\varepsilon_F/\Gamma)$ times the single-scattering rate and affects the mobility appreciably. The result is consistent with recent data [2].

2. Multiple Scattering by Ionized Impurities

In a uniformly doped system with a high impurity density, the electrons are predominantly scattered by the impurities at low temperatures. The matrix element for scattering by an ionized impurity of charge Ze at position $\vec{r} = 0$ is given, for a momentum transfer $\vec{q}$, by

$$U(\vec{q}) = -\frac{2\pi Ze^2}{S\kappa(q+s)}, \tag{1}$$

where $-e$, S, and κ are the elctronic charge, the area of the layer, and the bulk dielectric constant, respectively. The layer width is assumed to be much smaller than the Fermi wavelength. The screening constant is given in the RPA approxiamtion by $s = 2e^2m^*/(\kappa\hbar^2)$ for $q = |\vec{q}| \leq 2k_F$, where m^* is the effective mass and k_F the Fermi wave number. Only the lowest subband is assumed to be populated.

The single-impurity and two-impurity scattering processes are displayed in Fig.1a&b, respectively, in the standard diagram notations [1]. We find for the TRR for the single-impurity and two-impurity scattering processes (denoted by the superscripts 1 and 2, respectively)

$$1 / \tau_{ik}^{(1)} = \frac{2m^* S^2 N_i}{\hbar^3} \langle \sin^2\phi [U(2k\sin\phi)]^2 \rangle , \tag{2}$$

and

$$1 / \tau_{ik}^{(2)} = \frac{[U(0)]^2 [U(2k)]^2 \Gamma_{ik} \ell n \xi}{2\pi\varepsilon_k \langle [U(2k\sin\phi)]^2 \rangle \langle \sin^2\phi [U(2k\sin\phi)]^2 \rangle \tau_{ik}^{(1)}} , \tag{3}$$

where N_i is the impurity density per area and the angular brackets denote the angular average : $\langle F(\phi) \rangle = \frac{2}{\pi} \int_0^{\pi/2} F(\phi)\, d\phi$. In (3) $\xi = 4\varepsilon_k / \Gamma_{ik} \gg 1$ and Γ_{ik} is the damping of the state $\vec{k}$:

$$\Gamma_{ik} = \frac{S^2 m^* N_i}{2\hbar^2} \langle [U(2k\sin\phi)]^2 \rangle . \tag{4}$$

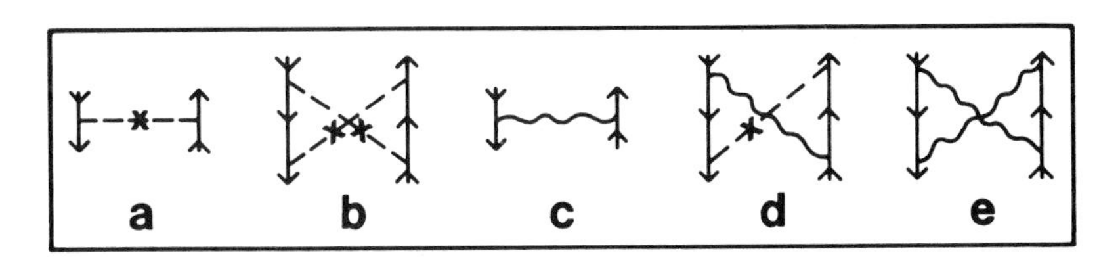

Figure 1. Contributions to the irreducible scattering part from impurity (dashed line with cross) and phonon scattering. Solid and wavy lines indicate Fermion and phonon propagators and the arrows the direction of the momentum and energy flow.

Note that $1/\tau_{ik}^{(2)}$ in (3) is not simply down by a factor Γ/ε_F compared to $1/\tau_{ik}^{(1)}$ but is also enhanced by the factor $\ell n\xi$. This logarithmic anomaly was first noted by Langer and Neal [3].

The mobility (μ) is determined from (2) and (3). At zero temperature it is of the form :

$$\mu^{-1} = aN_i + bN_i^2 \ell n(N_o/N_i) , \tag{5}$$

where $N_o = 4\varepsilon_F N_i/\Gamma$ ($>> N_i$) is independent of N_i in view of (4). The quantities a and b equal m^*/e times the coefficients of N_i and $N_i \ell n\xi$ on the right hand sides of (2) and (3) (cf. (4)), respectively.

We evaluate (5) for a GaAs quantum-well, using $m^* = 0.07\ m_o$ (m_o is the free electron mass) and $\kappa = 12$. In Fig.2 we exhibit the inverse mobility μ^{-1} in (5) for $N = 2.5\text{x}10^{11}$ [cm^{-2}] as a function of N_i in a solid curve. The dashed curve there corresponds to the single-impurity scattering contribution. In Fig.3, μ^{-1} is plotted for $N_i = 0.5\text{x}10^{11}$ [cm^{-2}] as a function of N^{-1} in a solid curve. The dashed curve there again corresponds to the single-impurity scattering contribution. The mobility is clearly affected significantly by two-impurity scattering. The coefficients a and b in (5) depend on energy and temperature ($\varepsilon \sim T$) as ε^{-1} and $\varepsilon^{-3}\ell n\varepsilon$ and decrease with increasing temperature. The density-(e.g., N, N_i) and temperature-dependent behaviors described

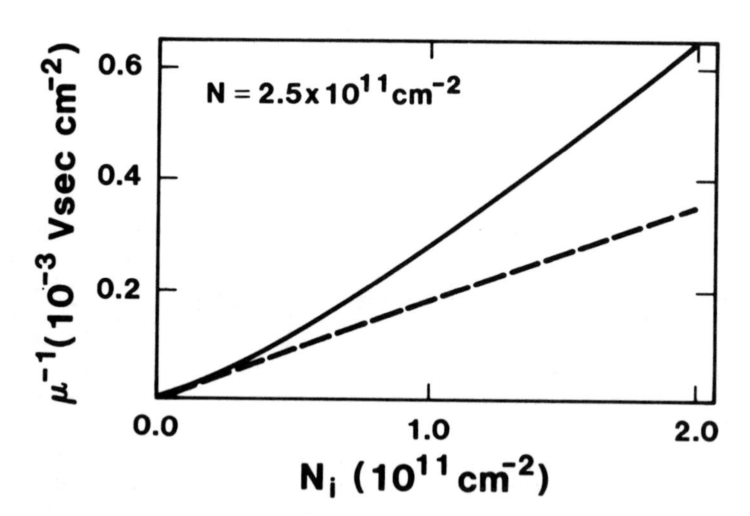

Figure 2. Inverse mobility in (5) as a function of N_i (solid curve). The dashed line represents the result of single impurity scattering.

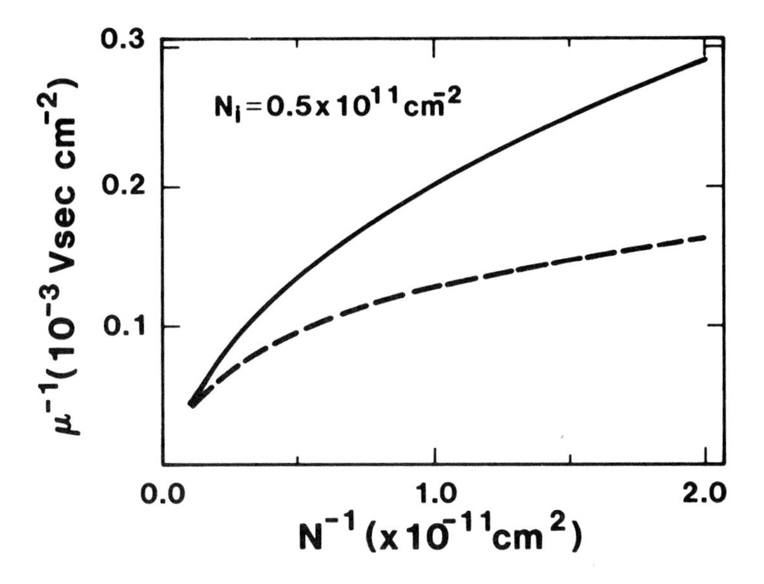

Figure 3. Inverse mobility in (5) as a function of N^{-1} (solid curve). The dashed curve represents the result of single impurity scattering.

above are very similar to those observed in Si-inversion layers [2] except that the data were analyzed in Refs. [2] and [4] according to (5) without the important logarithmic factor. This factor, for example, enhances the effect by a factor 3 for $\Gamma/\varepsilon_F = 0.2$ (corresponding to $N_i = 0.8\times10^{11}\,cm^{-2}$ in Fig.2).

3. Effect of Scattering by Acoustic Phonons

As the temperature increases, the ionized-impurity scattering becomes less efficient, while the scattering rate τ_p^{-1} by acoustic phonons increases linearly with temperature(T) [5]. The two rates become comparable at a certain temperature and the mobility reaches maximum. This crossover temperature depends on N_i, N, and the quantum-well width. The one-phonon scattering process is illustrated in Fig.1c. Multiple scattering processes corresponding to combined scattering by a phonon and an impurity as well as two-phonon scattering are shown in Fig.1d&e, respectively. In Fig.1d we should add a diagram obtained by interchanging the impurity and phonon lines. The processes in Fig.1d yield for the TRR

$$1 / \tau_{ipk}^{(2)} = \frac{2[U(2k)]^2 \Gamma_{ik} \ln\xi}{\pi\tau_p \langle [U(2k\sin\phi)]^2 \rangle \varepsilon_k} . \quad (6)$$

For the two-phonon scattering rate (Fig.1e) we find

$$1 / \tau_p^{(2)} = \frac{\Gamma_p \ln\xi}{\pi\tau_p \varepsilon_k} . \quad (7)$$

In (6) and (7) the subscripts i and p indicate impurity and phonon scattering and $\Gamma_p = \hbar\tau_p^{-1}$. The quantity Γ in $\xi = 4\varepsilon_k/\Gamma$ now includes contributions from impurity as well as phonon scattering. It is clear from our earlier discussion of the two-impurity scattering effect that the corrections in (6) and (7) will reduce the mobility significantly near the crossover temperature and above.

4. Concluding Remarks

In summary we have calculated the TRR to the order $(\Gamma/\varepsilon_F)\ln(4\varepsilon_F/\Gamma)$. Higher order $(\sim \Gamma/\varepsilon_F)$ contributions are expected from a damping correction to the electron propagators [4], from the $\Lambda_k(\varepsilon_F \pm i0, \varepsilon_F \pm i0)$-part of the vertex correction in Holstein's work [1], and also from other mechanisms. A detailed description of this work will be presented elsewhere.

Acknowledgement - This work was performed at Sandia National Laboratories supported by the U.S. Department of Energy under Contract Number DE-AC04-76DP00789.

References

1. T. Holstein, Ann. Phys. (NY) 29, 410 (1964).
2. A. Hartstein, A. B. Fowler, and M. Albert, Surf. Sci. 98, 181 (1980).
3. J. S. Langer and T. Neal, Phys. Rev. Lett. 16, 984 (1966).
4. S. Das Sarma, Phys. Rev. Lett. 50, 211 (1983).
5. P. J. Price, Ann. Phys. (NY) 133, 217 (1981).

Two Questions Raised by Simulations of the Hubbard Model

D.J. Scalapino

Abstract

Quantum Monte Carlo simulations of a quarter filled 1-D Hubbard model show that the onsite Coulomb interaction *suppresses* $2p_F$ charge density fluctuations. Simulations of the 2 and 3-D Hubbard model indicate that the p-wave pairing response is *suppressed* by U. These results suggest we take another look at our theoretical understanding of these many-body systems.

Results obtained from Monte Carlo simulations of interacting condensed matter systems can force us to take a second look at some of our ideas about many-body systems. Sometimes this second look merely reminds us of the limits inherent in certain approximations which over time we may have forgotten. Other times the numerical results suggest new physical features which have previously been neglected. Ted Holstein was always interested in the physics of a problem, and so here I would like to illustrate with two examples, how Monte Carlo simulations have led to a deeper physical understanding of the effect of an onsite Hubbard U interaction on the charge and spin correlations of a 1-D system and the effect of band structure on p-wave pairing in a 2 or 3-D Hubbard model.

The Hubbard model

$$H = -t\sum_{\langle ij\rangle}(c^{+}_{i\sigma}c_{j\sigma} + c^{+}_{j\sigma}c_{i\sigma}) + U\sum_{i} n_{i\uparrow}n_{d\downarrow} \quad , \tag{1}$$

in which electrons hop between near neighbor sites and have an onsite Coulomb interaction U, has been used to describe a variety of systems. For quasi one-dimensional materials, the properties of the 1-D Hubbard model have been extensively studied [1-3]. In one dimension, the non-interacting $U = 0$ system exhibits low temperature logarithmic divergences in its $2p_F$ charge and spin density susceptibilities as well as its zero momentum pairing susceptibilities. The interaction changes these to power laws, and the renormalization group [3] (RG) has provided a useful framework for analyzing their behavior. Here, for a non half-filled band where Umklap processes can be neglected, the interaction is described in terms of a large $2p_F$ momentum transfer coupling g_1 and a small momentum transfer coupling g_2. The resulting phase diagram obtained from the RG is shown in Fig. 1. The most divergent susceptibilities are used to characterize the different regions of the phase diagram. For the Hubbard model, $g_1 = g_2 = U/\pi v_F$ and according to Fig. 1, the low temperature phase is characterized by divergent $2p_F$ spin *and charge density* susceptibilities.

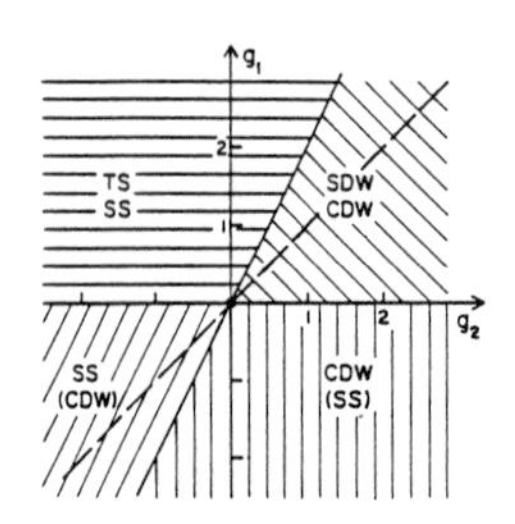

Fig. 1. Phase diagram of a one-dimensional electron gas obtained from the second-order renormalization group approximation. Coupling constants for the Hubbard model lie along the dashed line $g_1 = g_2$.

Using Monte Carlo simulation techniques we calculated the q-dependent spin and charge density susceptibilities [4]

$$\chi(q) = \int_0^\beta d\tau \, \frac{1}{N} \sum_l \langle (n_{i+l\uparrow}(\tau) - n_{i+l\downarrow}(\tau))(n_{i\uparrow}(0) + n_{i\downarrow}(0))\rangle e^{iql} \tag{2}$$

$$N(q) = \int_0^\beta d\tau \, \frac{1}{N} \sum_l \langle n_{i+l}(\tau) n_i(0)\rangle e^{iql} \tag{3}$$

for a one-quarter filled band. Plots of these susceptibilities for an inverse temperature $\beta = 7.25$ (in units where the bandwidth is 4) are shown in Fig. 2 for various values of U. At this temperature, the Monte Carlo data clearly indicate that the $2p_F$ charge-density susceptibility is suppressed by U in contrast with the RG expectation.

This prompted us to go back and look again at the RG formalism [4]. If U is repulsive, as in the present case, the effective renormalization couplings flow towards smaller values so that if $U/\pi v_F$ is small to start with, perturbation theory can be used to construct the RG. In second order one finds that [3]

$$\begin{aligned} \frac{dg_1(l)}{dl} &= -[g_1^2(l) + \tfrac{1}{2} g_1^3(l)] \\ g_2(l) &= g_2(0) - \frac{g_1(0)}{2} + \frac{g_1(l)}{2} \end{aligned} \tag{4}$$

Here $g_1(0)$ and $g_2(0)$ are the unrenormalized coupling constants of the original problem which for the Hubbard model are both equal to $U/\pi v_F$, and we have taken $l = \ln(4t/T)$. As the temperature is lowered and l becomes large, the effective couplings flow towards smaller values ending at the fixed point $g_1^* = 0$, $g_2^* = U/2\pi v_F$. A straightforward analysis of the $2p_F$ spin and charge density susceptibilities show that at the fixed point they both diverge as $(4t/T)^{\mu}$ with $\mu = 1 - \frac{(1-g_2^*)^{\frac{1}{2}}}{(1+g_2^*)^{\frac{1}{2}}} \simeq g_2^*$.

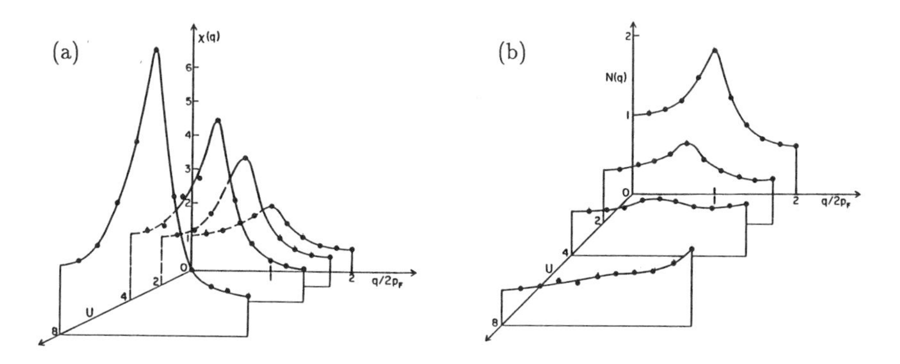

Fig. 2. (a) Spin-density and (b) charge-density susceptibilities at $\beta = 7.25$.

However, if one integrates the renormalization group equations and calculates the susceptibilities as a function of $l = \ln(4t/T)$, one obtains the results shown in Fig. 3a.

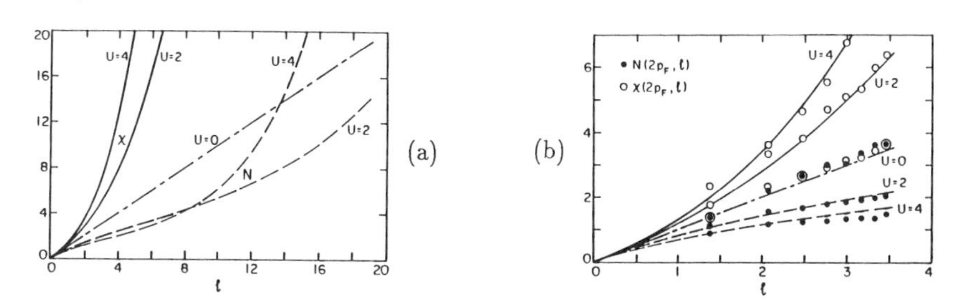

Fig. 3a,b. (a) Results for $N(2p_F)$ and $\chi(2p_F)$ versus $l = \ln(45/kT)$ obtained by integrating the second-order renormalization group equation, Eqs.(4), for different values of U. Note that initially the effect of U is to suppress N and enhance χ over the logarithmic divergence behavior of the free system shown by the dash-dotted line. Only at very large values of l corresponding to unphysically low temperature does the charge-density-wave susceptibility approach the same power-law divergences as the spin-density wave. (b) A blowup of the physically relevant temperature regime of (a). The solid points and open points are Monte Carlo data for $N(2p_F)$ and $\chi(2p_F)$, respectively, obtained for different values of U and different temperatures.

Although $N(2p_F)$ will indeed eventually diverge in the same way as $\chi(2p_F)$, this does not occur until l is very large. For a bandwidth 4t of order half an electron volt, an l value of 4 implies a temperature $T \sim 100K$, while an l value of 10 would correspond to a temperature less than $1K$. Thus the fixed point behavior is not relevant to the physical problem, and an onsite U acts to suppress the $2p_F$ site charge density fluctuations. As shown in Fig. 3b, the RG predictions in the region $l \leq 4$ are in excellent agreement with the Monte Carlo results.

The second example I will discuss involves the use of the 2 or 3-D Hubbard model to describe a system of fermions interacting through a short-range repulsive interaction. This system is expected to have spin fluctuation excitations and , based on simple models of ^3He, one might expect that the exchange of these spin-fluctuation excitations could lead to p-wave triplett pairing. The p-wave pairing susceptibility

$$P_p = \int_0^\beta d\tau \langle \Delta^{(p)}(\tau) \Delta^{(p)+}(0) \rangle \tag{5}$$

with

$$\Delta^{(p)} = \sum_p \sin p_x c_{p\uparrow} c_{-p\downarrow} \tag{6}$$

can also be written as

$$P_p = \int_0^\beta d\tau \sum_l \langle c_{l+i\uparrow}(\tau)\; c_{l+i+\hat{x}\uparrow}(\tau)\; c^+_{i+\hat{x}\uparrow}(0)\; c^+_{i\uparrow}(0) \rangle \tag{7}$$

Here $\hat{x}$ denotes a displacement of one lattice spacing in the x-direction. Just as the charge and spin density susceptibilities can be evaluated from a Monte Carlo simulation, so, too, can the p-wave pairing susceptibility. Unfortunately, presently available Monte Carlo techniques severely limit the lattice size and the lowest temperatures that can be achieved. Thus, HIRSCH [5] measured P_p on 2-D 6×6 lattices and on 3-D lattices $4 \times 4 \times 4$ lattices for $\beta = 3$. While one would like data on larger lattices at lower temperature, what was found clearly showed that turning on the interaction U *decreased* P_p rather than enhanced it. Again, this might simply mean that the lattices were too small and the temperatures too high, but it made us wonder whether the exchange of spin fluctuations on a lattice might give rise to a *repulsive* rather than an attractive p-wave interaction.

Looking at the Monte Carlo calculations one could see that even a small value of U led to a decrease in P_p. Thus it was natural to look at the lowest order perturbation theory contribution shown in Fig. 4. To this order, the effective interaction is given by

$$V_{eff}(p', p) = -\frac{U^2}{2}\left[\chi_0(p' - \vec{p}) - \chi_0(p' + \vec{p})\right] \tag{8}$$

with

$$\chi_0(q) = \sum_p \frac{f(\epsilon_{p+q}) - f(\epsilon_p)}{\epsilon_p - \epsilon_{p+q}} \tag{9}$$

For a free electron system χ_0 is the familiar Lindhard result shown in Fig. 5. Approximating it by the dashed curve

$$\chi_0 = N(0)(1 - aq^2) \tag{10}$$

with $a = 0.12$ gives

$$V_{eff}(p', p) = -2aN(0)U^2 \vec{p}\,' \cdot \vec{p} \tag{11}$$

This is clearly an attractive p-wave interaction. Summing the RPA paramagnon contributions enhances it by $(1 - N(0)U)^{-1}$. However, on a 3-D lattice, band structure effects lead to the possibility of Fermi surface nesting when the band is half-filled. This qualitatively changes the momentum dependence of $\chi_0(q)$, as shown in Fig. 6 for q along the [1,1,1] direction. For a range of μ values, $\chi_0(q)$ peaks at a finite value of $\vec{q}$ associated with momentum which span the nearly nested parts of the Fermi surface. Only for a nearly empty (or full) band, does $\chi_0(q)$ reduce to the Lindard result which peaks at $\vec{q} = 0$. The effect of this peak at finite q is to make the p-wave interaction repulsive. In a rough way one can say that the constant a in Eqs. (9) and (10) has changed sign.

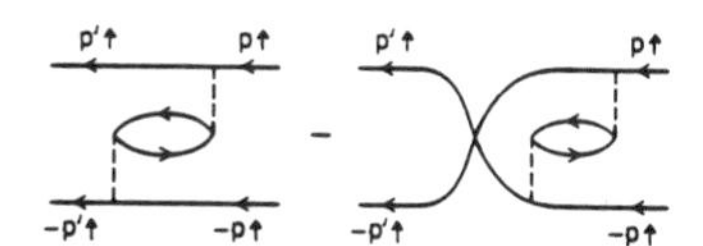

Fig. 4. The lowest order direct and exchange contributions to the effective triplet pairing interaction for the Hubbard model.

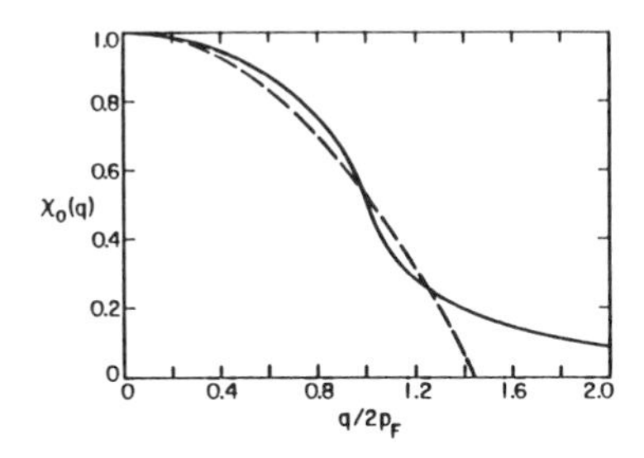

Fig. 5. The Lindhard $\chi_0(q)$ normalized to $N(0)$ versus $q/2p_F$. The dashed line is Eq. (10) with $a = 0.12$.

One might worry that this is a low momentum argument, so to see this more precisely [6], one can calculate the average p-wave coupling over the Fermi surface

$$\lambda_p = -\int \frac{dp^2}{v_p} \int \frac{d^2p'}{v_p'} \sin\, p_x'\; V(p', p) \sin p_x \Big/ \int \frac{d^2p}{v_p} \sin^2 p_x \tag{12}$$

with $V(p', p)$ the RPA paramagnon triplet interaction

$$V(p', p) = -\frac{U^2 \chi_0(p' - p)}{1 - U^2 \chi_0^2(p' - p)} \tag{13}$$

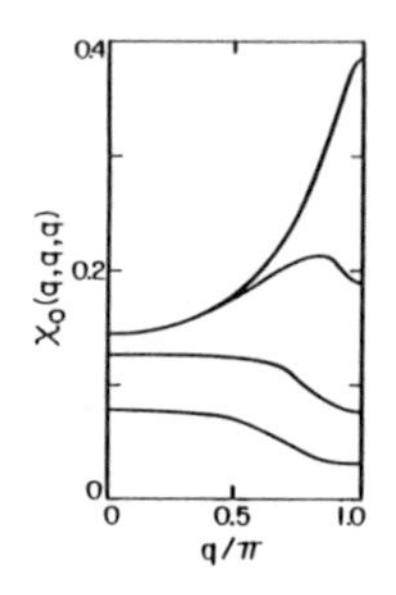

Fig. 6. $\chi_0(q, q, q)$ versus q for q along the [1,1,1] direction. The four curves were computed for $\beta = 4$ and correspond top to bottom to $\mu = 0, -1, -2$, and -3, respectively. As μ decreases χ_0 approaches the free electron behavior shown in Fig. 5.

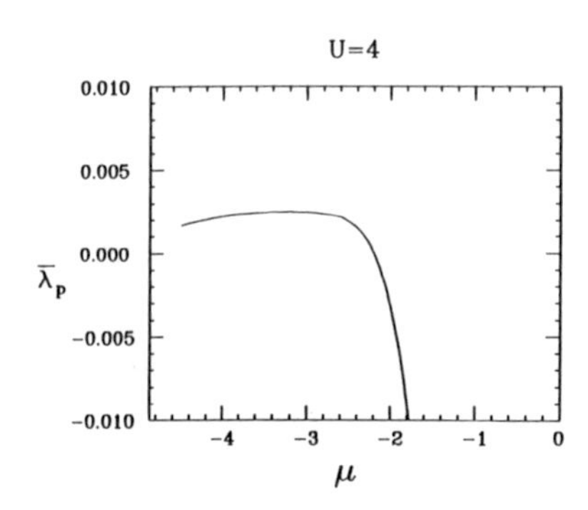

Fig. 7. The p-wave pairing coupling constant λ_p calculated from Eq. (12) versus μ. For small values of the band filling where the Fermi surface is spherical, λ_p is attractive but as μ increases and the band structure gives rise to nearly nested Fermi surface regions, λ_p becomes repulsive.

Figure 7 shows λ_p versus μ and clearly illustrates the way in which λ_p becomes repulsive as the band is filled. At low filling, $\mu < 3$, λ_p is indeed attractive as one expects from the continuum case. However, for a wide range of filling, band structure effects give rise to a repulsive p-wave coupling λ_p.

I like to think, that while Ted always questioned numerical approaches and worried that they would lead one to lose sight of the basic physics, he would look with more favor on their use as tests to make us take a more careful look at basic mechanisms. I do recognize, though, that he would most likely say we should have known it from the start.

Acknowledgment

This work was supported in part by the National Science Foundation under Grant DMR83-20481.

References

1. See, for example, *Recent Developments in Condensed Matter Physics*, edited by J.T. Devreese (Plenum, New York, 1981), Vol. 1, Chap. 8, and references therein.

2. V.J. Emery, in *Highly Conducting One-Dimensional Solids*, edited by J. Devreese, R. Evrard, and V. Van Doren (Plenum, New York, 1979), p. 247.

3. N. Menyhard and J. Solyom, *J. Low Temp. Phys.* **12**, 529 (1973); J. Solyom, *Adv. Phys.* **28**, 201 (1979).

4. J.E. Hirsch and D.J. Scalapino, *Phys. Rev. Lett.* **50**, 1168 (1983), and *Phys. Rev. B* **27**, 7169 (1983).

5. J.E. Hirsch, *Phys. Rev. B* **54**, 1317 (1985).

6. J.E. Hirsch, E. Loh, Jr. and D.J. Scalapino, to be published.

Polarons and Subsurface Bonding

Ivan K. Schuller and M. Lagos

Abstract

We have calculated the bonding energy of a hydrogen atom below the surface of Nb(110) and Pd(111) due to the interaction with the surface phonons. Our results show that at room temperature, there is a deep potential well just below the surface of Nb(110) with a much shallower well for Pd(111). Due to this, the kinetics of hydrogen absorption by Nb(110) surface and Pd covered Nb is drastically affected. The kinetic equations, modified to include this subsurface potential well, show that the subsurface-trapped hydrogen acts as a valve for the admission of hydrogen into the bulk. A large variety of experimental facts clearly follow from these considerations.

The work presented in this paper was done during a series of visits one of us (IKS) spent at the Catholic University in Santiago, Chile. This fact is related to Ted Holstein not only because it is about polarons but also because when our Chilean friends celebrate their day of independence (September 18) they also celebrate Ted's birthday. In fact, some years ago, we helped Ted celebrate with a bottle of good Chilean wine ("Marques de Casa Concha" 1975).

The behavior of hydrogen in metals is a field of active experimental and theoretical interests. In particular, the absorption of hydrogen by metallic surfaces has drawn much attention not only because of the possible applications, but also because a variety of interesting physical phenomena. In a series of fascinating experiments [1-3], it was shown that the absorption of hydrogen by the surface of niobium and tantalum is drastically modified by the coverage of three or more monolayers of palladium or platinum. This has been conventionally interpreted as due to a change in the electronic density of states at the Fermi surface, which presumably affects the dissociations of molecular hydrogen into atomic hydrogen [3]. We have proposed an alternate explanation which relies on the idea that in certain metals (for instance Nb) due to the interaction with the surface phonons, the hydrogen is bonded stronger below the surface than in the bulk ("subsurface bonding") [4]. In this fashion, the absorption of hydrogen is initially high, until the subsurface is saturated and then the surface further blocks the absorption into the bulk. The coverage of Nb with a few Pd monolayers reduce this subsurface binding and consequently the bulk absorption is enhanced. The subsurface bonding modifies the absorption kinetics in such a way that the subsurface coverage decrease abruptly at a critical temperature which depends on the bulk concentration [5].

The energy of a hydrogen atom in an interstitial site can be calculated using the standard second quantized Hamiltonian formalism [6]. Since the calculation relates to surface properties both the bulk and surface phonon contributions ought to be taken into account. Neglecting the terms that describe diffusion (i.e., low temperatures) and assuming that the lattice-mediated interactions between the hydrogens is small (low concentrations) the hydrogen energy was shown to be given by [4]

$$E = \sum_{\lambda} m\varepsilon_0 - m \frac{|g_\lambda|^2}{\hbar\omega_\lambda} \tag{1}$$

where m is the mass of hydrogen atom, $m\varepsilon_0$ is the energy of the impurities in a rigid lattice, ω_λ is the frequency of vibrational mode λ and g_λ is the normalized λ-th Fourier component of the "force" between the hydrogen atom and the host ion and therefore the second term describes the self trapping energy.

In ref. [4] we have calculated explicitly g_λ and performed numerical calculations to evaluate the value of the self trapping energy under some simplifying assumptions.

The surface contributions to the self trapping energy for layer n (= 0,1,2,...) was found

$$\Delta_s(n) = \frac{1}{8\pi^2 M(N_s/S)U^2} \int_0^{2Q_D} dx\, \frac{1}{x}\, e^{-2h(\ell_z/a)x}\, A_x^2 \int_0^{2\pi} d\phi |F(x,\phi)|^2 \tag{2}$$

where a is the lattice parameter, (N_s/S) is the number of metal atoms per unit surface area, M is the mass of the metal ion, the vibrational modes are assumed to be Debye like i.e., $\omega_\sigma = UQ$ for $Q<Q_D$ and $\omega_\rho = vq$ for $q<q_D$, $\ell_z \equiv na_z$, $h \equiv [1-(U/v)^2]^{1/2}$, $Q_D = (4\pi N_s/S)^{1/2}$, $A_x^2 = 1 - \exp(-2ha_z x/a)$ and

$$F(x,\phi) = \sum_{\vec{\ell}} \exp(i\vec{Q}\cdot\vec{L} - hQ\ell_z)\ F(\vec{\ell})\ \hat{e}_\sigma \cdot \hat{\ell} \qquad (3)$$

with $\vec{\ell} = (\vec{L},\ell_z)$ a vector going from the interstitial site $\vec{\ell}$ to an ionic lattice site. The parameter $F(\vec{\ell}) \equiv |\langle\nabla V(\vec{r}-\vec{\ell})\rangle|$ is the force the hydrogen impurity exerts on the ℓth lattice ion and depends on lattice symmetry and surface orientation.

In spirit the present calculation is similar to the now famous Holstein polaron [7] with the difference that ours is a "hydrogenic" as opposed to "electronic", polaron and that we explicitly calculate the contribution of the surface terms to the energy.

We have calculated the total solution energy at site n by adding the experimentally measured bulk solution energy E_{sol} to the self trapping energy, $\Delta_s(n)$. The parameters used in the calculation are shown in Table I.

Table I

Parameters Used in the Calculation

	$-E_{sol}$ (eV)	M (10^{-22}g)	N_s/S ($1/a^2$)	a (Å)	v (10^5cm/s)	$-E_{ads}$ (eV)	F (10^{-4} dyn)
Nb(110)	0.358[8]	1.543	$\sqrt{2}$	3.30	2.419	0.55[10]	3.035
Pd(111)	0.200[9]	1.765	$4/\sqrt{3}$	3.89	2.255	0.45[11]	1.263

F is obtained for (Nb, tetrahedral) and Pd (fcc, octahedral) from spectroscopic measurements [12,13] of the bulk lattice distorsion caused by the interstitial hydrogen and theoretical calculation of ref. [4].

The surface speed of sound U, has not been determined experimental so we performed calculations using both the upper and lower theoretical limits ($0.87 \leq U/v \leq 0.95$). The total solution energy are given in Table II.

Table II

Total Solution Energies for Various Depths

	Tetrahedral site in Nb(110)	Octahedral site in Pd(111)
n	$-E_n$(eV)	$-E_n$(eV)
0	0.562 - 0.808	0.253 - 0.337
1	0.427 - 0.461	0.220 - 0.234
2	0.385 - 0.393	0.208 - 0.212
3	0.371 - 0.374	

Recently, experimental evidence [14,15] has been found which is claimed to prove conclusively the existence of subsurface bonding in accordance with our theoretical ideas.

From a comparison of Table I and II it is clear that subsurface bonding is more important for Nb than Pd because E_0 (subsurface bonding energy) > E_{ads} (chemisorption energy) for Nb and $E_0 < E_{ads}$ for Pd. As a consequence the absorption kinetics of Nb and Pd will be quite different [5]. In the case of Nb, the initial uptake will be large until the subsurface is saturated and then the H saturated surface, blocks the bulk from further absorption. Since the subsurface self trapping is much smaller in Pd the bulk uptake in Pd (or Pd covered Nb) will be much higher.

In order to calculate the temperature dependence of the surface coverage θ_b we have modified [5] the kinetic equations originally written down by CONRAD, ERTL and LATTA [11,16] to include the deep subsurface bonding. In this calculation the site to site transfer rate perpendicular to the surface is taken to be activated. Assuming a quasistationary state the kinetic differential equations can be solved and the dependence of surface coverage can be calculated relatively easily. Figure 1 shows the

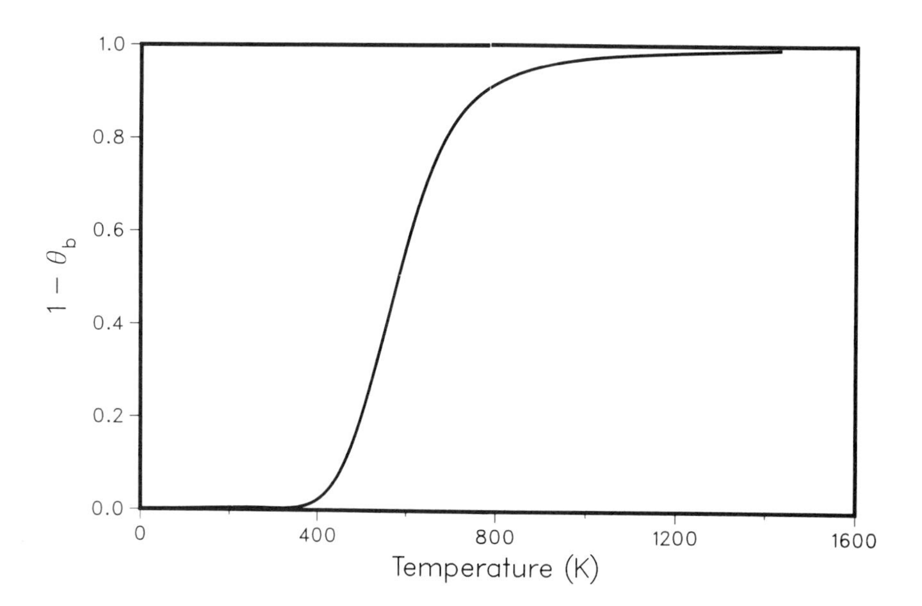

Fig. 1. Temperature dependence of the $1-\theta_b$ for a critical temperature T_c = 600 K, where θ_b is the coverage of the subsurface layer.

dependence of $1-\theta_b$ on temperature with reasonable choices of parameters for the case of Nb.

It is easy to understand the experimental hydrogen absorption by Nb(110). At low temperature, $\theta_b \sim 1$, the subsurface becomes saturated fast initially and then the surface and bulk become decoupled, as observed [10]. A thin (~ 3 monolayers) overlayer of Pd decreases the depth of the subsurface well bringing it closer to the bulk interstitial energy. In this fashion, the surface does not block the passage of hydrogen into the bulk and therefore the uptake increases. At this point, we would like to stress that the existence of the subsurface bonding mechanisms is shown by our calculations and that the qualitative conclusions are independent of any reasonable variation of the parameters. Moreover, it is clear that the existence of a deep subsurface well has to affect the kinetics of hydrogen absorption in a fundamental way [17].

Our theory also predicts several new phenomena which can be tested experimentally. In particular, the strong temperature dependence of the subsurface coverage θ_b has not yet been studied experimentally.

To summarize, a calculation of the interaction of hydrogen with surface phonons shows the existence of a deep potential well under the subsurface of certain transition metals. A variety of experiments have confirmed the existence of the subsurface bonded hydrogen predicted here. As a result of the subsurface bonding the kinetics of absorption is modified so that the bulk and surface become completely decoupled. This calculation is in agreement with a variety of experimental observations related to the absorption of hydrogen by the Nb(110) and Pd covered Nb(110) surfaces.

One of us (IKS) has benefited from fruitful discussions and a long lasting friendship with Beverlee and Ted Holstein. Work supported by the U.S. Department of Energy, BES-Materials Sciences, under Contract #W-31-109-ENG-38 and the Organization of American States (OAS).

1. M. A. Pick, J. W. Davenport, M. Strongin and G. J. Dienes: Phys. Rev. Letters 43, 286 (1979).
2. M. Strongin, M. El-Batanouny and M. A. Pick: Phys. Rev. B22, 3126 (1980).
3. M. El-Batanouny, M. Strongin, G. P. Williams and J. Colbert: Phys. Rev. Letters 46, 269 (1981).
4. M. Lagos and I. K. Schuller: Surf. Scie. Letters 138, L161 (1984).

5. M. Lagos, G. Martinez and I. K. Schuller: Phys. Rev. B29, 5979 (1984).
6. C. Kittel: Quantum Theory of Solids, (J. Wiley and Sons eds., New York 1983), pg. 16.
7. T. Holstein: Ann. Phys. (NY) 8, 325, 343 (1959).
8. J. A. Pryde and C. G. Titcomb: J. Phys. C5, 1293 (1972).
9. G. Comsa, R. David and B. J. Schumacher: Surf. Scie. 95, L210 (1980).
10. R. J. Smith: Phys. Rev. B21, 3131 (1980).
11. H. Conrad, G. Ertl and E. E. Latta: Surf. Scie. 41, 435 (1974).
12. H. Pfeiffer and H. Peisl: Phys. Letters 60A, 363 (1977).
13. J. M. Rowe, J. J. Rush, H. G. Smith, M. Mostoller and H. E. Flotow: Phys. Rev. Letters 33, 1297 (1974).
14. R. J. Behm, V. Penka, M. G. Cattania, K. Christman and G. Ertl: J. Chem. Phys. 78, 7486 (1983).
15. K. H. Rieder, M. Baumberger and W. Stocker: Phys. Rev. Letters 51, 1799 (1983).
16. M. A. Pick: In Metal Hydrides, ed. by G. Bambadakis (Plenum, New York 1981).
17. The solution of the kinetic equations has been criticized by a comparison of experimental data and our theoretical calculation (G. J. Dienes, M. Strongin and O. Welch: Phys. Rev. B32, 5478 (1985)). This criticism has been shown to be incorrect (M. Lagos and I. K. Schuller: Phys. Rev. B 32, 5477 (1985)).

New Directions in Calculating Electron-Phonon Interactions

Marvin L. Cohen

Abstract

A discussion of the role of recent advances for calculating electron-phonon coupling constants is given. The focus is on an *ab initio* self-consistent pseudopotential approach which requires only the atomic number and atomic mass as material parameters.

1. Introduction

An accurate description of the electron-phonon (EP) interaction is central to understanding many properties of solids. This interaction is responsible for electrical resistivity and its disappearance in the superconducting state. It produces polarons, dispersion in phonon spectra, and to some extent influences many, if not most, properties of materials. From a fundamental theoretical view, a description of the EP interaction can be an excellent tool for studying basic physical interactions between particles and fields since the process describes the interaction of a fermion with a scaler boson field.

Ted Holstein understood both the basic nature and the practical aspects of EP couplings and applied this knowledge to unravel the role of the coupling in polaron properties. At this Symposium, Bob Schrieffer and David Emin have discussed Holstein's contributions in this area, and several individuals have described his passion for a rigorous approach to problems without unnecessary empirical input. Despite Holstein's strong feelings about avoiding experimental input, in the early 1960's when I discussed the possibility of superconductivity in degenerate semiconductors with him, he agreed that the use of experimentally determined EP coupling constants was necessary and appropriate. A first-principles approach wasn't possible at that time. However, a great deal of progress has been made in the development of *ab initio* methods to compute material properties, and recently a first-principles scheme for EP calculations has been developed. One success is the calculation of EP couplings in highly compressed silicon which allowed a successful prediction [1] of superconductivity in this material. Although Ted Holstein would have preferred a completely analytic treatment of the problem, rather than a computer calculation of the relevant couplings, one consolation is that the only experimental input to this calculation was the atomic number, the atomic mass, and the crystal structure.

2. Electron-Phonon Interactions

The rigid-ion approximation has been one of the most successful models for calculating EP coupling constants. This approach is well-documented [2], and the EP part of the hamiltonian is given by,

$$H_{ep} = \sum_{la} \vec{\delta}R_l^a \cdot \vec{\nabla}V_a(\vec{r}-\vec{R}_l^a) \quad . \tag{1}$$

The displacement $\vec{\delta}R_l^a$ for cell l and atom a is expanded in phonon coordinates [2], and the atomic potential V_a can be represented by a pseudopotential [3]. In the rigid ion approach, the potential at each lattice site $\vec{R}_l^a$ is assumed to move rigidly, and the electron readjustment is not included. Empirical pseudopotentials [3] can be used for V_a to obtain the wavefunctions for computing the EP matrix elements [4,5].

A more direct and basic approach is based on the total-energy pseudopotential scheme [6] in which pseudopotentials are generated from atomic wavefunctions and electron-electron interactions are calculated using the local density approximation [7]. The calculation of the total energy for different crystal structures and for "frozen-in" phonon distortions allows a determination of the structure and the phonon spectrum. The extension to EP coupling [8] involves the evaluation of the self-consistent change of the crystal potential caused by a phonon distortion of amplitude $u_{\vec{q}\nu}$ where $\vec{q}$ and ν represent the phonon wavevector and branch. The EP matrix element [2] depends on the quantity $\hat{\varepsilon} \cdot \vec{\nabla}V \equiv (V_{\vec{q}\nu} - V_0)/u_{\vec{q}\nu}$ where $V_{\vec{q}\nu}$ and V_0 are the self-consistent potentials of the distorted and undistorted crystal, respectively, and $\hat{\varepsilon}$ is the polarization vector [9]. The EP matrix element [2] becomes

$$g(\vec{q},\nu) = \left(\frac{\hbar}{2M\omega}\right)^{1/2} \delta(\vec{k}-\vec{k}'-\vec{q}) \langle n,\vec{k}|\hat{\varepsilon}\cdot\vec{\nabla}V|n'\vec{k}'\rangle \tag{2}$$

where the $\langle n,\vec{k}|$ wavefunctions represent Bloch electron states, M is the atomic mass, and ω is the phonon frequency. For metallic systems, a $\vec{q}$-dependent EP parameter can be defined

$$\lambda_\nu(\vec{q}) = 2N(E_F)\,\frac{\langle\langle|g(\vec{q},\nu)|^2\rangle\rangle}{\hbar\omega} \tag{3}$$

where $\langle\langle\rangle\rangle$ represents an average over the Fermi surface and $N(E_F)$ is the density of states at the Fermi energy. The standard McMillan [10] EP parameter λ is given by

$$\lambda = \frac{1}{\Omega_{BZ}} \int \sum_\nu \lambda_\nu(\vec{q})\, d^3q \tag{4}$$

where Ω_{BZ} is the Brillouin zone volume. A calculation of the phonon linewidth [11] can also be performed directly

$$\gamma_\nu(\vec{q}) = \pi N(E_F)\lambda_\nu(\vec{q}) \quad . \tag{5}$$

The above scheme was applied to Al [8] where the electronic band structure, phonon dispersion curves, EP $\vec{q}$-dependent couplings and linewidths for longitudinal and transverse acoustic modes, and the parameter λ were all computed from first principles. The results were found to be consistent with the available experimental data.

3. Superconductivity

Three standard ways [2] of measuring the EP couplings in a metal are the electronic heat capacity, the high temperature electrical resistivity, and the superconducting transition temperature, T_c. Of these, T_c depends most sensitively on λ [10]

$$T_c = \frac{\Theta_D}{1.45} \exp\left[\frac{-1.04(1+\lambda)}{\lambda-\mu^* - 0.62\lambda\mu^*}\right] \tag{6}$$

where Θ_D is the Debye temperature and μ^* is the repulsive electron-electron Coulomb parameter. The μ^* parameter is difficult to compute from first principles, but estimates are available [10,12]. Hence with computed values of λ and Θ_D, it is possible to estimate T_C, and this was done for two highly compressed hexagonal phases of Si.

Silicon is known to transform from the diamond to a white tin metallic phase around 100 kbar. This phase is superconducting [13] at 6.3 K. Total energy-pseudopotential calculations [14] explored other possible phases and indicated that a hcp metallic form of Si would be stable in the 400 kbar range. Two experimental groups [15,16] discovered this phase and another simple hexagonal (sh) phase at lower pressures ~130 kbar. Calculations [17] of the properties of the sh phase revealed that this material had a large $N(E_F)$ and a covalent-like charge distribution. Both of these features are conducive for superconductivity [18], and an experimental search was suggested. However, it was a direct calculation using the methods described here which finally motivated the experimental discovery [1]. Theory predicted a $T_C \sim$ 5-10 K and a decreasing dependence of T_C on pressure (P). The measurements yielded a value of 8.2 K at 152 kbar which is one of the highest T_C's for a nontransition elemental metal. The measured $T_C(P)$ curve [1] has negative slope and decreases to 3.6 K at 250 kbar.

More detailed calculations [19,20] suggested that the $T_C(P)$ curve would reach a minimum and then rise near the sh → hcp phase transition and that the hcp phase would be superconducting. Preliminary measurements [21] appear to verify both predictions, but some structure is seen in the 360 kbar range where Olijnyk et al. [16] suggest the possible existence of a new phase of Si at pressures between the sh and hcp transition pressures. Recent measurements on Ge verify the prediction of a white tin → sh transition in the 800 kbar range [22]. However, this work indicates that there is a transition from sh → double hcp around 1.1 Mbar. This suggests the possibility of a similar transition in Si between sh and hcp, but at this time, the question of the existence of another structural phase in Si remains unclear.

4. Conclusions and Future Research

A serious limitation on the calculations of the EP λ is the inclusion of enough points for $\lambda_\nu(\vec{q})$ ((3)) for the averaging in (4). The calculations described above used only the [001] phonons branches and coupling to these phonons. Calculations for other directions, [011] and [010], are being done, but progress is slow because of the large amounts of computer time required. The calculated pressure dependence, $\lambda(P)$ [19,20], was based only on the [001] modes to estimated $T_C(P)$ after T_C was fit at one pressure by adjusting μ^* in (6). Hopefully, future calculations will include a better estimate of λ and T_C by computing more $\lambda_\nu(\vec{q})$.

A related area which should be explored in greater detail is the connection between the structural phase transitions in compressed Si and T_C. In the analysis of the [001] mode, it was found that the EP coupling increased for the TA mode near the sh-hcp transition, and this phonon mode drives the transition. A detailed experimental and theoretical analysis of transitions of this kind may clarify the nature of the relation between changes in T_C and structural phase changes. Phenomena of this kind have been observed in many systems, but a general theoretical description of their properties is not available.

Another area of research is the search for new materials with large EP interactions for high temperature superconductivity or other applications. It is interesting to speculate that metastable phases may be the most interesting systems to study. Although the total-energy pseudopotential approach can yield the properties of these phases, at present the procedure is to use a trial structure as input and to compare its structural energy with other structures. However, in the future it may be possible to use statistical methods [23] to find trial structures and to explore the possibility of their stability or metastability. The properties of possible phases can be calculated, and the most interesting ones suggested for experimental study using diamond anvil techniques to reach high pressures.

This new area of predicting the existence of high pressure phases of solids and their properties such as superconductivity appears to be promising, and the calculation of EP couplings is an important aspect of the research. However, better formal and computational techniques and critical analyses on the "Holstein" level are required to make the approaches described here of practical use in material science. At this point, it's only fair to say that the "new directions" in this area seem to have promise.

5. Acknowledgements

This work was supported by National Science Foundation Grant No. DMR83-19024 and by the Director, Office of Energy Research, Office of Basic Energy Sciences, Materials Sciences Division of the U.S. Department of Energy under Contract No. DE-AC03-76SF00098.

6. References

1. K. J. Chang, M. M. Dacorogna, M. L. Cohen, J. M. Mignot, G. Chouteau, G. Martinez: Phys. Rev. Lett. 54, 2375 (1985).
2. G. Grimvall: Electron-Phonon Interaction in Metals, ed. by E. P. Wohlfarth (North-Holland, Amsterdam, 1981).
3. M. L. Cohen and V. Heine: Solid State Phys. 24, 37 (1970).
4. M. L. Cohen, Y. W. Tsang: J. Chem. Phys. Solids 32, (Suppl. 1), 303 (1971).
5. O. J. Glembocki, F. H. Pollak: Phys. Rev. B 25, 1193 (1982).
6. M. L. Cohen: Physica Scripta T1, 5 (1982).
7. W. Kohn, L. J. Sham: Phys. Rev. 140, A1333 (1965).
8. M. M. Dacorogna, M. L. Cohen, P. K. Lam: Phys. Rev. Lett. 55, 837 (1985).
9. This description is appropriate for one atom per cell, but is is easily generalized to a larger basis.
10. W. L. McMillan: Phys. Rev. 167, 331 (1968).
11. P. B. Allen: Phys. Rev. B 6, 2577 (1972); P. B. Allen, B. Mikovic: Solid State Phys. 32, 1 (1982).
12. K. H. Bennemann, J. W. Garland: in Superconductivity in d- and f-Band Metals, ed. D. H. Douglass, AIP Conference Proceedings No. 4 (American Institute of Physics, New York, 1972), p. 103.
13. J. Wittig: Z. Phys. 195, 215 (1966).
14. M. T. Yin, M. L. Cohen: Phys. Rev. Lett. 45, 1004 (1980).
15. J. Z. Hu, I. L. Spain: Solid State Comm. 51, 263 (1984).
16. H. Olijnyk, S. K. Sikka, W. B. Holapfel: Phys. Lett. 103A, 137 (1984).
17. K. J. Chang, M. L. Cohen: Phys. Rev. B 30, 5376 (1984); 31, 7819 (1985).
18. M. L. Cohen, P. W. Anderson: in Superconductivity in d- and f-Band Metals, ed. D. H. Douglass, AIP Conference Proceedings No. 4 (American Institute of Physics, New York, 1972), p. 17.
19. M. M. Dacorogna, K. J. Chang, M. L. Cohen: Phys. Rev. B 32,

1853 (1985).

20. M. L. Cohen, K. J. Chang, M. M. Dacorogna: Physica 135B, 229 (1985).
21. D. Erskine, P. Y. Yu: Bull. Am. Phys. Soc. 31, 640 (1986).
22. Y. K. Vohra, S. Desgrenier, A. L. Ruoff, K. J. Chang, M. L. Cohen: Phys. Rev. Lett. (in press).
23. R. Car, M. Parrinell: Phys. Rev. Lett. 55, 2471 (1985).

The Electron-Phonon Cornucopia

J. Robert Schrieffer

1. Introduction

Systems consisting of electrons coupled to lattice vibrations or phonons constitute some of the simplest yet richest dynamical structures in condensed matter physics. Ted Holstein made fundamental contributions to these problems, as many of the speakers in this symposium will describe. These problems include the structure and dynamics of polarons, charge transport in metals, localization, charge density waves, *etc.* In each area Ted sought to answer questions which could be clearly stated and which have fundamental significance for the understanding of real world problems. His standards were uncommonly high, standards which he applied equally to his own work and to that of others. Nevertheless he was a scientific realist, pursuing specific results, often taking years of focussed effort. His style and his humanity have made us all stand a little taller.

2. The Polaron

The simplest of these problems is that of one electron imbedded in a uniform elastic medium, socalled jellium, since the medium can vibrate but has no underlying crystal lattice structure. Crudely, the problem is similar to the behavior of a billiard ball placed on a stretched elastic sheet. The sheet sinks in the vicinity of the ball to lower the potential energy of the system, forming a potential well. The depth of the well depends on the stiffness of the sheet, shallow wells occurring for stiff sheets and deep (large binding energy) wells for soft sheets. As the ball moves slowly, the well follows, gradually losing its front to back symmetry at high velocity, when inertia of the sheet (finite velocity of sound) becomes important. We might call this moving entity, ball plus elastic deformation surrounding it an elasteron. As the speed of the elasteron increases, the back side of the deformation will not in general return to the original undeformed sheet. Rather, sound waves will be left behind in a wake, leading to frictional drag. Finally, if the sheet is held at a finite temperature, the elasteron will be buffetted by thermally excited elastic deformations and undergo diffusive motion.

In the corresponding polaron problem the Hamiltonian is

$$H = \sum_k \epsilon_k n_k + \sum_q \hbar\omega_q N_q + g \sum_{k,q} F_q C^+_{k+q} C_k (b_q + b^+_{-q}) \tag{1}$$

where $\epsilon_k = \hbar^2 k^2/2m, \omega_q$ is the phonon energy, n_k and N_q are the electron and phonon occupation numbers $C^\dagger$ and $b^\dagger$ create an electron and a phonon, while g is the coupling constant and F_q the form factor. As David Emin will discuss, the electron localization and elastic energies oppose the electron-phonon coupling energy. When the coupling constant g is large, the electron digs a deep hole and "self traps," taking on a very large effective polaron mass $m_p >> m$. For small g the electron-phonon effects are spread out in space.

Ted studied in great detail the strong polaron, particularly in the physically relevant case where the electrons and phonons live on a lattice, such as in a molecular crystal. In this case, the strong polaron can sit on any site, with identical average energy. Because of quantum fluctuations the elastic deformation surrounding a neighboring site can be momentarily equal to that of the original site and the electron can hop, leaving the original site either deformed (inelastic) or undeformed (elastic). This strong polaron hopping conductivity is the complement of the weak coupling transport studied by Felix Block[1] in his celebrated paper of 1929.

Thus, the simplest of all electron-phonon problems, the polaron, is enormously rich, exhibiting quasi phase transitions (large to small polaron), temperature active conductivity and temperature induced resistivity, resonances in the frequency dependent transport, *etc.*. Again, our understanding of many of these phenomena was greatly advanced by Ted's insights.

3. <u>The Normal Metal</u>

While the polaron model well describes semiconductors and insulators, where the mean spacing between electrons (or holes) is large compared to the size of a given polaron, metals fall in the opposite limit. While in the jellium model of a metal the Hamiltonian is again given by (1), two complications arise, namely 1) the phase space for electron motion is reduced by the Pauli principle and 2) a phonon emitted by one electron can be absorbed by another, *i.e.*, there are phonon mediated electron electron interactions. These effects lead to superconductivity in metals.

The normal (non superconducting) phase of the metallic electron phonon system was first properly treated by A.B. Migdal [1,2] using the Green's function approach of quantum field theory. He showed that as a consequence of the smallness of the ratio of the speed of sound and the fermi velocity $c_S/v_F < 10^{-2}$ in typical metals, the one-electron Green's function $G(k)$ could be calculated to the same accuracy $\sim 10^{-2}$, even if the coupling constant g is greater than unity. In essence, he proved that the electron-phonon vertex function $\Gamma(k,q)$ is unity, with corrections of order the adiabatic parameter c_S/v_F. Therefore, $G(k)$ when expressed in terms of an energy-momentum vector $k = (k_o, \mathbf{k})$

$$G(k) = \frac{1}{k_o - \epsilon_k - \sum(k)} \tag{2}$$

has a self-energy $\sum(k)$ given by

$$\sum(k) = -g^2 \int G(k+q) D(q) \frac{d^4q}{(2\pi)4}. \tag{3}$$

While (2) and (3) form a coupled set, Migdal proved that to the same accuracy, G in (3) can be replaced by the zero order electron's Green's function defined by (2) with $\sum(k)$ set equal to zero. Thus, the coupled electron phonon system is reduced to quadrature to leading order in the adiabatic parameter.

The electron spectral function

$$A(k) = \frac{1}{\pi} |ImA(k)| \tag{4}$$

shows a quasi particle peak centered at $E_k = \frac{m}{m*}\epsilon_k$, where $\epsilon_{k_F} \equiv 0$ and $\frac{m*}{m}$ is the mass enhancement ratio,

$$\frac{m*}{m} = \left(1 + \frac{\partial \sum}{\partial \epsilon_k}\right) \Big/ \left(1 - \frac{\partial \sum}{\partial k_o}\right) \tag{5}$$

where $k = k_F$ and $k_o = 0$. Typically $m*/m \sim 1.2 - 2.2$ in metals and one is not in the small polaron limit.

A long standing question in the theory of transport in metals is why one can use the weak coupling Baltzmann equation with the electron-phonon collision integral calculated to lowest order (g^2) using zero order fermi and Bose distribution functions? In particular, if the width γ_k of the quasi particles is large compared to k_BT, is it not necessary to include complicated higher order effects? Ted, amongst others, showed that as a consequence of Σ being short range in space but long range in time, all such corrections cancel out, again to leading order in c_S/v_F. This is a highly significant result, justifying the approximations of the original Block paper, even for rapidly decaying quasi particles.

4. Superconductivity

Since Migdal solved the electron-phonon problem in metals to leading order in c_S/v_F and did not find superconductivity, what was missed? Superconductivity is

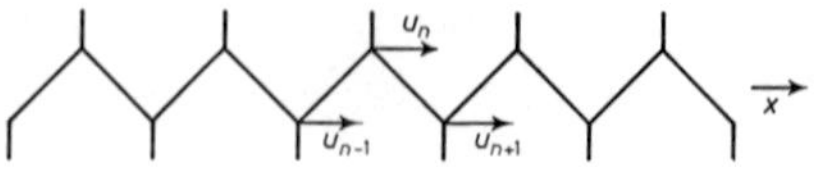

Figure 1a

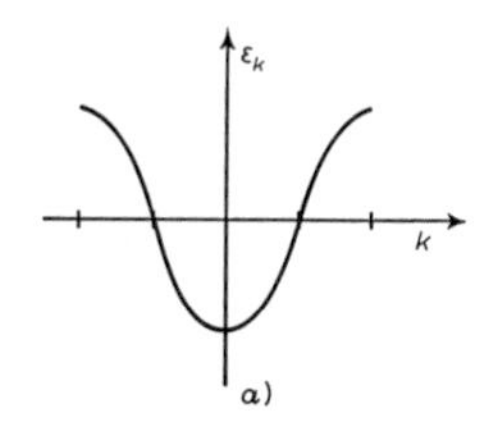

Figure 1b

intrinsically a nonperturbative phenomenon arising from a condensation of electron pairs into a highly correlated state [2]. Formally, the difficulty with Migdal's approach arises from singularities in the perturbation expansion of the vertex function

from carbon $2p_z$ atomic orbitals. Since this band can accommodate two electrons per carbon, the band is half full. Naively, one would expect $(CH)_x$ to be a metal, like sodium with no gap at the fermi surface, as shown in Fig. 1b. However, as Peierls pointed out, a one dimensional metal is unstable with respect to lattice distortion of wavenumber $Q = 2k_F = \pi/a$ regardless how weak the electron phonon coupling happens to be. (Here a is the fundamental period of the undistorted crystal.) This is because nesting is perfect in an ideal one dimensional system. The situation is illustrated in Fig. 2. In Fig. 2a, the spacing between carbons joined by a "double bond" is shortened somewhat (by about 0.04 Å in practice) while the spacing of carbons joined by a single bond stretches by a corresponding amount. While the chemists' pictures used here suggest that the double bond is of full strength, *i.e.*, contains one electron of each spin as opposed to $\frac{1}{2}$ an electron of each spin in the undistorted lattice, the actual π electron density is only weakly modulated, being $\sim 1.2, 0.8, 1.2, 0.8$ *etc.* rather than $2, 0, 2, 0, \ldots$. Thus, one has a charge density wave of small amplitude. Nevertheless, this modulation leads to a gap 2Δ in the electron spectrum which is of order 1.4 eV, compared to the π band width of 10 eV.

By symmetry, it is clear that the inverse distortion shown in Fig. 2b leads to the same energy as that in Fig. 2a. We denote these inequivalent states A and B. The electronic structure associated with A and B are identical, and as shown in Fig. 2c the filled states are lowered in energy by the distortion. The total energy continues to decrease in either A or B until the increase of the elastic energy of stretched σ

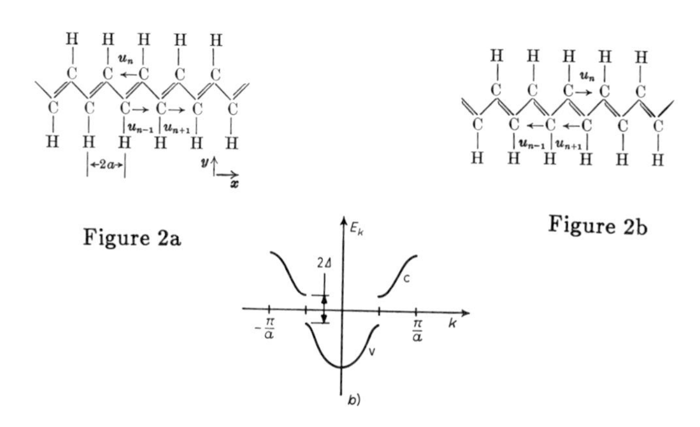

Figure 2c

bonds balances the electronic energy lowering. By definition this occurs when the magnitude of the displacement of each carbon is u_0.

Within the mean field approximation, the displacements of the carbons are fixed, independent of the presence of excited electrons or holes. In this approximation the system is a conventional semiconductor with gap 2Δ. However, as Su, Heeger and the author showed, these excitations are unstable with respect to the formation of solitons, S, and anti solitons, $\bar{S}$, the actual elementary excitation of the system.

S and $\bar{S}$ are illustrated in Fig. 3. S is a domain boundary separating on the left and A on the right, while $\bar{S}$ is the opposite. The electronic structure of S is very simple. For each soliton, a state ψ_o, localize about S, is split off into the gap. For widely spaced solitons, the energy of ψ_o is near the center of the gap. Since the system has particle-hole symmetry and total state count is conserved, it follows that the conduction and valence bands are each depleted by one half a state per spin orientation. If the state ψ_o is unoccupied, the missing valence band state density $(\frac{1}{2}) \times 2$ spins $= 1$ state leads to a net charge $+e$ and spin zero. This is, however, the reverse spin-charge relation compared to electrons or holes in metals and semiconductors, where elementary excitations of charge $\pm e$ carry spin $\frac{1}{2}$. Similarly, if ψ_o carries one electron, the soliton charge is zero, but the spin is one half. Again, this is the reverse spin charge relation compared to excitons which carry charge zero and spin zero or unity.

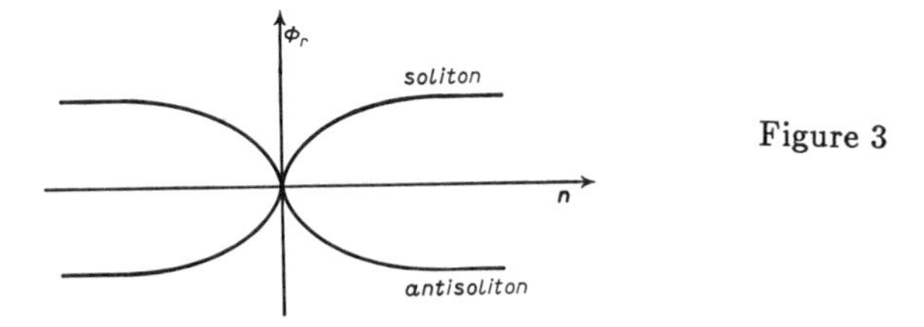

Figure 3

Roughly speaking, we can view a soliton as half of a polaron. This satisfies the requirement that a polaron has two electronic states per spin in the gap arising from the symmetric and antisymmetric combinations of the gap orbitals ψ_o and $\bar{\psi}_o$ of S and $\bar{S}$. Alternatively, a polaron is a bound state of S and $\bar{S}$. The reverse spin charge properties have been observed experimentally, as have many other properties of the soliton model of these broken symmetry chain conductors.

One remarkable prediction of the soliton model is that these excitations in general carry sharply defined fractional charge. For example, for a one third filled band, solitons are predicted to carry charge $\pm\frac{2}{3}e$ and spin zero or $\pm\frac{1}{3}e$ and spin $\frac{1}{2}$. $(CH)_x$ soliton would carry charge $\pm\frac{e}{2}$ were it not for the two fold spin degeneracy. Solitons carrying fermion number $\pm\frac{1}{2}$ were first predicted by Jackiw and Rebbe in the context of a one space dimensional relativistic quantum field theory model. Subsequently, such peculiar excitations have been predicted in many different situations, including the three space dimensions.

6. Conclusion

The coupled electron phonon system is one of the simplest and richest models in condensed matter and quantum field theory. From a single Hamiltonian flows such diverse phenomena as polaron formation, weak to strong polaron transition, fermi liquid behavior in dense electron systems, phonon mediated superconductivity, charge density wave formation, dynamic symmetry breaking, soliton excitations and fractional charge. Undoubtedly, the list continues with many new phenomena remaining to be discovered. Ted Holstein played a unique role in this remarkable field, establishing standards of physical understanding and formal elegance which will ensure that this field will serve as a model for theorists in decades to come.

References

1. G.D. Mahan, *Particle Physics*, Plenum (1981).
2. J.R. Schrieffer, *Theory of Superconductivity*, W.A. Benjamin, Inc. N.Y. (1983).
3. W.P. Su, J.R. Schrieffer and A.J. Heeger, *Phys. Rev. Lett.* **42**, 1698 (1979); *Phys. Rev.* **B22**, 2099 (1980).
4. J.R. Shrieffer, "Highlights of Condensed Matter Theory," Course LXXXIX (Varenna Summer School) Italian Physical Society, Bologna, Italy pp. 300–348.
5. A.J. Heeger, S. Kivelson, J.R. Schrieffer and W.P. Su, *Rev. Mod. Phys.*, to be published.
6. R. Jackiw and C. Rebbi, *Phys. Rev.* **D13**, 3398 (1976).

Magnetic Interaction in a 2-D Electron Gas

D. Shoenberg

1. Introduction

When I came to the University of Pittsburgh for a sabbatical half year in 1962, I had just completed a study of the de Haas-van Alphen (dHvA) effect in copper, silver and gold [1]. The main purpose of the study was to determine the Fermi surfaces (FS) of the metals, but I also found some peculiar features of the dHvA oscillations which could be plausibly interpreted if it was supposed that the electrons 'perceived' the magnetic field as B rather than H. Ordinarily, the difference $4\pi M$ between B and H is too small to make any appreciable difference, but if the oscillations are sufficiently strong to make $4\pi dM/dB \sim 1$, there is a kind of feed-back effect, which has come to be known as magnetic interaction (MI) and which can significantly modify the form of the oscillations. I consulted Ted Holstein about this problem and asked him if it was possible to prove by microscopic theory that the relevant field was indeed B rather than H. After quite a short while he told me that within the Hartree approximation he could give such a proof, but he had some reservations and it was only 11 years later that he published a detailed analysis in collaboration with NORTON and PINCUS [2]. In the meantime PIPPARD [3] had given a thermodynamic argument to show that B rather than H was the relevant field. Since its discovery over 20 years ago MI has been found to have a variety of interesting consequences and many of these have been demonstrated experimentally; a detailed account is available in two recent reviews [4,5].

During the last few years there has been considerable interest in the magnetic behaviour of essentially 2-D rather than 3-D electron systems and the purpose of this paper is to review possible effects of MI in such systems; some aspects of this problem have been considered previously

[6,7,8]. In idealized conditions the oscillation cycles of magnetization in such 2-D systems are triangular rather than sinusoidal and the problem of MI becomes particularly simple. Two kinds of system will be considered: (1) a 'genuine' 2-D system such as can be produced in the inversion layer of a MOSFET by a gate voltage or in a heterojunction, and (2) a metal, such as a suitably intercalated graphite, with a cylindrical or nearly cylindrical FS; such a 'quasi' 2-D system can show much stronger dHvA oscillations than a genuine 2-D system and so the possibility of MI is more relevant.

2. Theoretical Results Ignoring Magnetic Interaction

The theory of the magnetization oscillations of a 2-D electron gas was first given by PEIERLS [9] and has been recently reviewed by SHOENBERG [10] with emphasis on the effects of finite T, broadening of Landau levels, sample inhomogeneity and electron spin. Here we shall ignore these complications and quote the results for 'ideal' conditions: T = 0, sharp Landau levels, perfect sample, no spin. The results also assume that the electrons can be treated as independent particles. With these assumptions the energies of the Landau levels are given by

$$\varepsilon_r = (r + \tfrac{1}{2})\beta B, \qquad \text{where} \tag{1}$$

$$\beta = e\hbar/mc \tag{2}$$

and m is the cyclotron mass. Each such level has a D-fold degeneracy and so can be occupied by at most D electrons, where

$$D = eB/\pi\hbar c \tag{3}$$

The magnetic moment M of the system is given by $-\partial E/\partial B$, where E is the total energy, if the number N of particles is independent of field, but if it is ζ, the chemical potential (the Fermi energy if electrical effects can be ignored) which is independent of field, M is given by $-\partial\Omega/\partial B$, where Ω is the thermodynamic potential defined as $E - N\zeta$. For a parabolic band, with m and β independent of r and B in (1), the necessary summations to obtain E or Ω are simple, and explicit formulae for E, Ω and M are easily obtained [5,10]. These formulae still apply to a good approximation even if the band is non-parabolic, provided the quantum number n, the value of r for the highest occupied level, is large; the relevant value of m is then that at the top of the band. If ζ remains constant as B is varied, all the levels below ζ are completely full and those above completely empty. As B increases, the levels rise smoothly until the highest filled level passes through ζ, when it empties suddenly, producing a cusp in the variation of Ω with B and a jump in M. For the range of a single cycle of oscillations specified by

$$n - \frac{1}{2} < F/B < n + \frac{1}{2}, \quad \text{we find} \tag{4}$$

$$\Omega = N_0\beta\left(\frac{1}{2} n^2 \frac{B^2}{F} - nB\right) \tag{5}$$

$$M = -\partial\Omega/\partial B = N_0\beta\left(n - n^2 \frac{B}{F}\right), \tag{6}$$

where F is the dHvA frequency given by

$$F = \zeta/\beta \tag{7}$$

and N_0 is the number of particles in the absence of a field. The actual number N of particles has a field variation about N_0 similar to the variation of -M. The various parameters have slightly different meanings according as the system is quasi or genuinely 2-D.

For the quasi 2-D system

$$\zeta = N_0\pi\hbar^2 d/m \quad \text{and} \quad F = N_0\pi\hbar c d/e \tag{8}$$

where $2\pi/d$ is the length of the cylindrical FS between zone planes, i.e. d is the unit cell dimension along the cylinder axis and N_0 is now to be taken as per unit volume. With this definition of N_0, M in (6) is also per unit volume and is therefore the magnetization. For the genuine 2-D system

$$\zeta = \nu_0\pi\hbar^2/m \quad \text{and } F = \nu_0\pi\hbar c/e \tag{9}$$

where ν_0 is the number per unit area. If N_0 in (6) is replaced by ν_0, the left hand side of (6) becomes μ, the magnetic moment per unit area. It is perhaps plausible to ascribe a thickness t to the 2-D layer, where t is of the order of the extent of the electron wave function normal to its plane; typically $t \sim 3 \times 10^{-7}$ cm. It is then possible to estimate a magnetic moment M per unit volume, i.e. a magnetization such that

$$M \simeq \mu/t \tag{10}$$

If it is N rather than ζ that remains constant as B varies, the Fermi energy 'sticks' to the highest occupied level, and the occupation of this level falls as B increases, while all the lower levels remain completely full. As soon as the highest level has emptied, ζ jumps to the next lower level which then begins to empty. The range of a single oscillation is now defined by

$$n \ < \ F/B \ < \ n + 1 \tag{11}$$

and in this range

$$E \ = \ N\beta[(n + \tfrac{1}{2})B - \tfrac{1}{2}\, n(n + 1)\, \frac{B^2}{F}] \quad \text{and} \tag{12}$$

$$M \ = \ -\partial E/\partial B \ = \ N\beta[n(n + 1)\, \frac{B}{F} - (n + \tfrac{1}{2})], \quad \text{where} \tag{13}$$

$$F \ = \ \zeta_0/\beta \tag{14}$$

and ζ_0 is the Fermi energy in the absence of a field; the actual Fermi energy ζ has a field variation around ζ_0 similar to the variation of M. Just as before, the interpretation of these equations is different for the quasi and the genuine systems. For the former

$$\zeta_0 \ = \ N\pi\hbar^2 d/m \qquad \text{and} \qquad F \ = \ N\pi\hbar cd/e, \tag{15}$$

where N is per unit volume and d is as defined above; with this definition M in (13) is again the magnetization. For the latter

$$\zeta_0 \ = \ \nu\pi\hbar^2/m \qquad \text{and} \qquad F \ = \ \nu\pi\hbar c/e \tag{16}$$

where ν is the number per unit area. Again, if N in (13) is replaced by ν the left hand side of (13) becomes μ the magnetic moment per unit area and the magnetization M is given by (10).

In order to be able to consider MI it is convenient to plot $4\pi M$ as a function of B and we obtain the triangular oscillations of Figs 1 and 2 corresponding to (6) and (13) respectively. It should be noted that the oscillations for N constant are almost exactly a mirror image of those for ζ constant, as reflected about the B axis, apart from a phase shift of approximately $\frac{1}{2}$ a cycle. The slopes $-\gamma'$ and γ of the inclined portions of the cycle in the two diagrams are given for the quasi 2-D system by

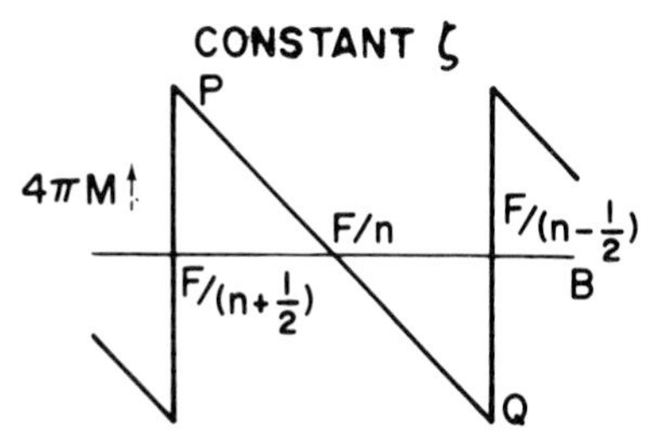

Fig. 1 $4\pi M$ vs B; slope of PQ = $-\gamma'$.

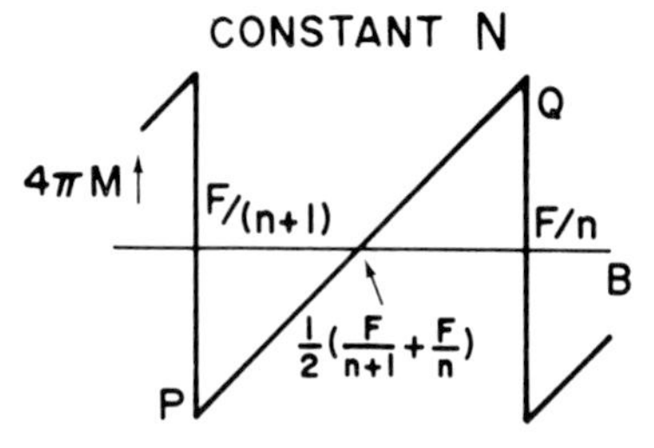

Fig. 2 $4\pi M$ vs B; slope of PQ = γ.

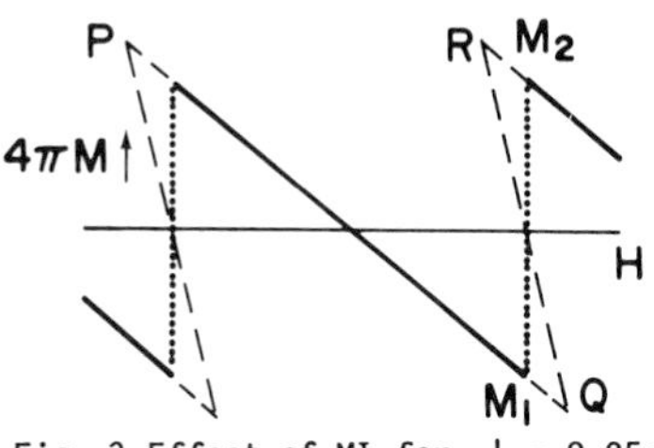

Fig. 3 Effect of MI for $\gamma' = 0.25$; $4\pi M$ vs H; slope of PQ = $-\gamma'/(1+\gamma')$ = -0.2, of QR = -1.

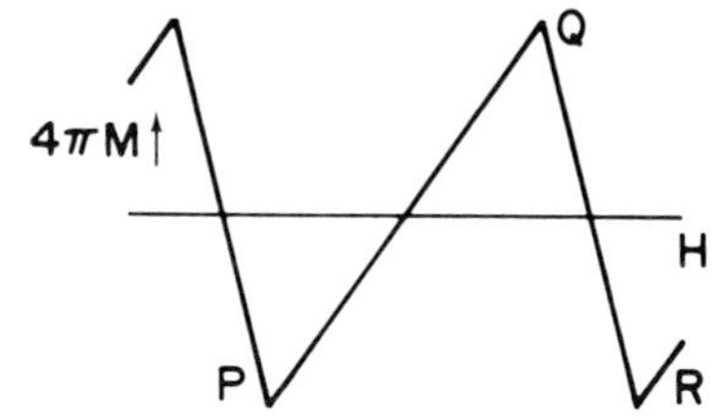

Fig. 4 Effect of MI for $\gamma = 0.25$; $4\pi M$ vs H; slope of PQ = $\gamma/(1-\gamma) = 1/3$, of QR = -1.

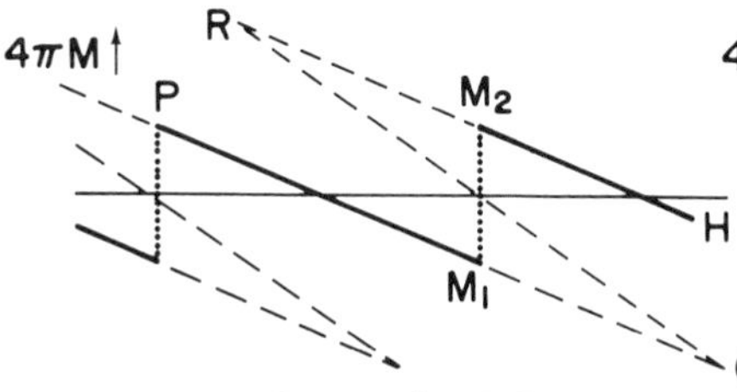

Fig. 5 As Fig. 3, but $\gamma' = 1.5$; slope of PQ = -0.6, of QR = -1.

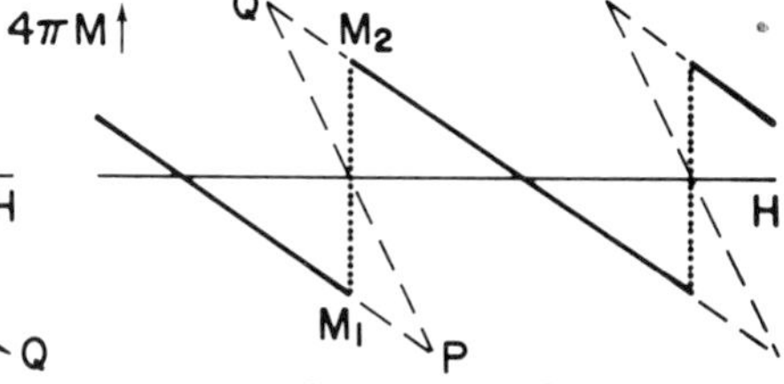

Fig. 6 As Fig. 4, but $\gamma = 1.5$: slope of PQ = -3, of QR = -1.

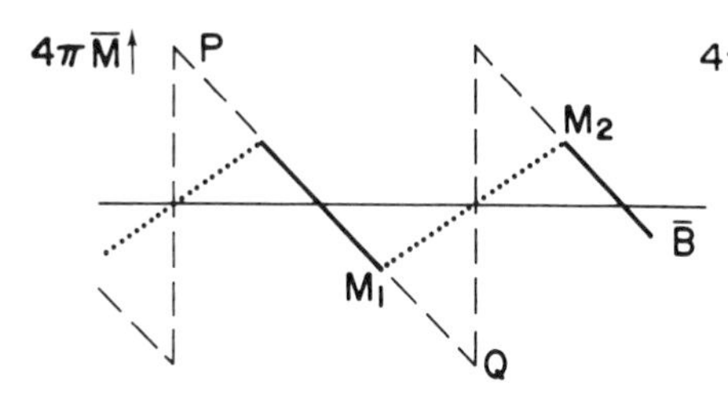

Fig. 7 As Fig. 5 but $4\pi\bar{M}$ vs $\bar{B}$; slope of PQ = $-\gamma' = -1.5$, of M_1M_2 = 1.

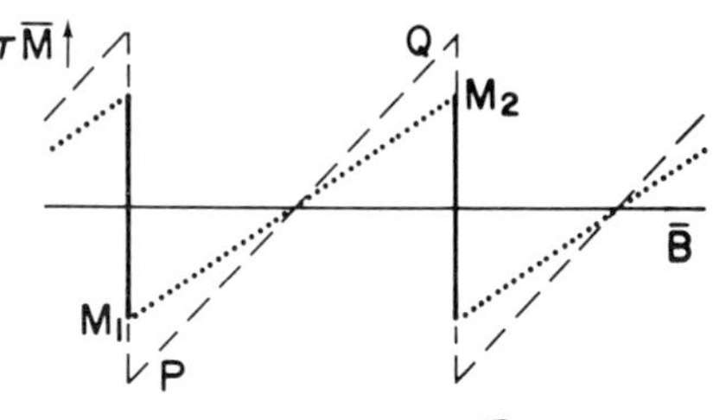

Fig. 8 As Fig. 6 but $4\pi\bar{M}$ vs $\bar{B}$; slope of PQ = $\gamma = 1.5$, of M_1M_2 = 1.

<u>Notes</u> Figs 1, 3, 5, 7 for constant ζ, Figs 2, 4, 6, 8 for constant N. In each series M = 0 occurs at the fields indicated in Figs 1, 2. The scale of ordinates is the same for Figs 5 to 8, but expanded 6-fold in Figs 3, 4 and arbitrary in Figs 1, 2. Dashed lines indicate metastable or unstable states; dotted lines show the ranges of break up into domains of magnetizations M_1 and M_2.

$$\gamma' = 4e^2n^2/mc^2d \qquad \text{for } \zeta \text{ constant, and} \tag{17}$$

$$\gamma = 4e^2n(n+1)/mc^2d \qquad \text{for N constant,} \tag{18}$$

while for the genuine 2-D system d must be replaced by t. It should be note that for $n \gg 1$, $n \simeq F/B$, so

$$\gamma' \simeq \gamma \simeq (4e^2/mc^2d)(F/B)^2 \tag{19}$$

where B is a field in the range of the cycle considered.

The effect of non-ideal conditions (finite T, imperfect sample, electron scattering and electron spin) has been discussed in a previous paper [10] an the discussion will not be repeated here. It should however be borne in min that the non-ideality will round the sharp corners of the oscillations and reduce their amplitude, particularly at low fields.

3. The Effect of Magnetic Interaction

In Figs 1 and 2 it is immaterial whether the field variable is B (as shown) or H, since MI has been ignored. If, however, $4\pi dM/dB$ is not neglibibly small the difference cannot be ignored and the graph of $4\pi M$ against H become significantly different from that against B; H is of course essentially the field determined by the magnet if the sample has zero demagnetizing coefficient. Since

$$H = B - 4\pi M \tag{20}$$

and $4\pi M$ varies linearly with B, the effect of MI is particularly easily determined, in contrast to the 3-D situation for which the variation is approximately sinusoidal. Thus if we rewrite (6) and (13) over the respective ranges (4) and (11) as

$$4\pi M = -\gamma'(B - F/n), \text{ and} \tag{21}$$

$$4\pi M = \gamma[B - F(n + \tfrac{1}{2})/n(n+1)] \tag{22}$$

for ζ constant and N constant respectively, we find using (20)

$$4\pi M = \frac{-\gamma'}{1 + \gamma'} (H - F/n), \text{ and} \tag{23}$$

$$4\pi M = \frac{\gamma}{1 - \gamma} [H - F(n + \tfrac{1}{2})/n(n + 1)] \tag{24}$$

The vertical sections of the cycles, i.e. B = constant, become simply

$$4\pi M = -H + C \tag{25}$$

where C is $F/(n + \frac{1}{2})$ and $F/(n - \frac{1}{2})$ for the ζ constant cycle and $F/(n + 1)$ and F/n for the constant N cycle.

The cycles are illustrated in Figs 3 to 6 for γ' and $\gamma < 1$ and > 1. It can be seen that for ζ constant there is always a range of field around $F/(n + \frac{1}{2})$, $F/(n - \frac{1}{2})$, etc. for which M is multivalued: this range is small for $\gamma' \ll 1$ but increases in proportion to γ'. For N constant, M becomes multivalued only if $\gamma \geqslant 1$ and then over a range around $F(n + \frac{1}{2})/n(n + 1)$ proportional to $(\gamma - 1)$. As was first pointed out by PIPPARD [3], parts of these multivalued regions are unstable or metastable and the equilibrium configuration is that corresponding to the smallest value of Ω or of E. This argument shows that for certain discrete values of H there are discontinuities in M as shown in Figs 3, 5, 6; at these points the two extreme values of M can coexist, but intermediate local values of M are unstable. It can be seen that the effect of MI for ζ constant is to reduce the amplitude of the cycle by $(1 + \gamma')$ without altering its shape, whatever the magnitude of γ'. For N constant, however, the effect of MI is more drastic with the shape of the cycle changing progressively as γ increases until for $\gamma \geqslant 1$ the discontinuities in M set in and the cycle assumes the form of Fig. 6 independent of γ. We may note that the arguments leading to the discontinuities could equally well have been presented in terms of H vs B graphs in close analogy to the p vs V graph for a liquid-gas transition; the instabilities come about because dH/dB cannot be negative.

So far it has been assumed that the sample is of zero demagnetizing coefficient so that H is the same as H_e, the field of the magnet. The case of a sample with finite demagnetizing coefficient D was first discussed by CONDON [11] who showed that over certain ranges of H_e the sample should break up into domains of differing M. Provided the division into domains when it occurs is on a sufficiently small scale,

$$H_e = H + 4\pi D\bar{M} = \bar{B} - 4\pi(1 - D)\bar{M} \tag{26}$$

where $\bar{M}$ and $\bar{B}$ are the means of M and B through the sample (for the ranges c H_e over which there is no domain formation $\bar{M}$ and $\bar{B}$ are, of course, the same as M and B). For simplicity we shall consider only the limiting case of a thin plate normal to the field, for which D = 1; the arguments are easily generalized to other values of D if required, but the results are somewhat less transparent. For D = 1 then, (26) becomes

$$H_e \quad = \quad H + 4\pi\bar{M} \quad = \quad \bar{B} \tag{27}$$

and the graphs of $4\pi\bar{M}$ against H_e are identical with those of $4\pi M$ against B, except that the jumps of M at values H = C in Figs 4 to 6 become translate into linear sections of slope 1, i.e.

$$4\pi\bar{M} \quad = \quad H_e - C, \tag{28}$$

in the graphs of $4\pi\bar{M}$ against H_e. Over these ranges the sample breaks up ir a mixture of domains having the constant magnetizations M_1 and M_2 indicatec As H_e increases, the proportion of M_2 domains increases from 0 to 1 while that of the M_1 domain decreases from 1 to 0, and $\bar{M}$ increases correspondingl

It is important to note that H has the same value throughout the sample over the domain range but the local M and B change quasi-discontinuously fr one domain to the next. Although $\bar{B}$ can be varied continuously, the local B within each domain can assume only discrete values and as emphasized by PIPPARD [4], certain ranges of B are completely excised by MI. Another interesting point is that for a 3-D metal in the limit of 'strong' MI the oscillations assume the same triangular form as for $\gamma \geqslant 1$ in the 2-D consta N situation. The 2-D constant ζ graph also approaches this form in the lim of $\gamma' >> 1$.

The existence of domains in a 3-D metal has been demonstrated by CONDON and WALSTEDT [12] and the nature of the domain structure has been discussed by CONDON [11] and PRIVOROTSKII [13]. This discussion shows that for a pla of thickness Z and cyclotron orbit radius R, the domain width X is given by

$$X \quad = \quad \alpha(ZR)^{1/2} \tag{29}$$

where α is a dimensionless constant of order unity. This estimate is, however, valid only provided that the domain width X is large compared to t thickness of the domain wall, which is of order R. Thus the theory must break down for $Z \lesssim R$. What happens then is a problem not yet fully resolved either experimentally or theoretically, but it has been suggested [14,15] that perhaps the structure is similar to that of a superconducting film normal to the field, when flux penetrates the film in single flux quanta. In the present context, of course, these vortices would carry only the local differences of B across the sample.

4. Orders of Magnitude and Relevant Experiments

4.1 Genuine 2-D

Typical values for a MOSFET or for a GaAs/AlGaAs heterojunction are:

$$\nu \sim 5 \times 10^{11} \text{ cm}^{-2} \qquad m \sim 0.2\, m_0, \text{ so}$$

$$F \sim 10^5 \text{ G} \qquad |\mu| \sim 2 \times 10^{-8} \text{ Gcm.}$$

Here m_0 is the free electron mass and the mod sign denotes half the peak-to-peak excursion of the oscillations. For $B = 5 \times 10^4$ G, we find

$$\gamma' \sim \gamma \sim 10^{-4}$$

and MI should not occur (except for negligibly small ranges of fields if ζ rather than N is constant).

Experimental observation of the magnetization oscillations by modulation methods (such as attempted by FANG and STILES [16]) is difficult not only because the magnitudes involved are so small, but because it is difficult to be certain that the effect of eddy currents is completely negligible. A more successful approach has been to measure the magnetic moment by the torque acting on the sample when the field is inclined to the normal. EISENSTEIN et al. [17] enhanced the torque considerably by using 50 identical layers of heterojunction closely packed together and the observed oscillations (Fig. 9) agree extremely well with the ideal oscillations of Fig. 2 if modified by a Gaussian broadening of the Landau levels with the Gaussian width varying as $B^{1/2}$. The sense of the asymmetry of the oscillations points definitely to the constant N rather than the constant ζ situation.

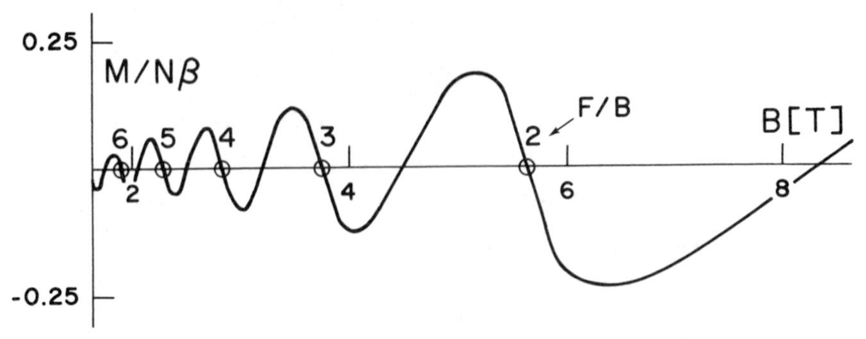

Fig. 9 Oscillations of M in a sample of 50 layers of GaAs/AlGaAs Heterojunction (after EISENSTEIN et al. [17]) $T = 0.4$ K, $\nu = 5.4 \times 10^{11}$ cm^{-2}, implying $F = 11.2$ T, which agrees well with the zeros of M as indicated.

4.2 Quasi 2-D

Usually the FS of intercalated graphites contains more than one almost cylindrical sheet and the true situation is more complicated than the ideal situations envisaged above. In reality it is the sum of the numbers of carriers in the various sheets which is constant as B is varied and moreover there may be MI between the oscillations of the various sheets [5]. Some idea of the relevant orders of magnitude may however be obtained by ignoring these complications and treating each cylindrical FS as if it was independent. In certain circumstances, however, the presence of another and larger part of the FS may effectively provide a reservoir of carriers so that it becomes more appropriate to think of ζ rather than N as constant when B is varied.

For the stage 2 Br intercalated graphite sample studied by MARKIEWICZ et al. [8], dHvA oscillations were observed corresponding to light and heavy holes and approximate relevant values for the corresponding sheets of FS (assumed exactly cylindrical) are

$$F_1 \sim 2.5 \times 10^6 \text{ G}, \qquad m_1 \sim 0.1\, m_0$$

$$F_2 \sim 11 \times 10^6 \text{ G}, \qquad m_2 \sim 0.2\, m_0 .$$

The relevant unit cell dimension is $d \sim 10^{-7}$ cm, so we find from (15) $N_1 \sim 1.2 \times 10^{20}$ cm^{-3}, $N_2 \sim 5.5 \times 10^{20}$ cm^{-3} and from (19) for $B = 5 \times 10^4$ G

$\gamma_1 \sim 0.3, \qquad |M_1| \sim 10$ G

$\gamma_2 \sim 3, \qquad |M_2| \sim 25$ G

From (19) it would seem that γ could be greatly increased by going to lower fields, but in practice non-ideal conditions would rapidly take over in reducing γ as B was lowered (e.g. at T = 1 K the low frequency amplitude at 2×10^4 G would be roughly halved and at 4.2 K reduced by a factor 25 or so).

Since no account has been taken of the detailed band structure, these estimates are inevitably rough, but it is clear that γ could well be of order unity or larger, even after allowing for some reduction due to non-ideal conditions, so that MI could be relevant. The reason that γ is so much larger for the quasi than for the genuine 2-D case is of course basically that γ varies as F^2 and because of the higher electron density, F is much higher for the quasi 2-D case. It should also be remembered that $|M|$ is per unit volume and the 'effective volume' of a genuine 2-D sample - even for 50 layers - is very small compared with that of a practicable quasi sample.

Studies of the dHvA effect in stage 2 Br intercalated graphite by MARKIEWICZ et al. [8] using the field modulation technique do indeed show striking features in the lower frequency oscillations even at 4.2 K (where the higher frequency is almost completely damped out because of its higher cyclotron mass); some of these features are interpreted by the authors as evidence for MI and in particular for domain formation.

5. Discussion

The experiments of EISENSTEIN et al. [17] on the oscillations in a genuine 2-D sample are of particular interest not only in giving the form of the Landau level broadening, but also in demonstrating that the form of the oscillations seems to be characteristic of constant N rather than ζ. The Quantum Hall Effect (QHE), however, suggests that over the plateaus of the Hall voltage the effective N must vary, as if ζ was constant with varying B, and this is supposed to be a consequence of the localized nature of the states in the 'wings' of the Landau level broadening. It therefore seems puzzling that samples similar to those used in the dHvA study showed a 'well-developed' QHE, as stated in [17]. Perhaps 'well-developed' may not imply that the plateau occupied more than a small fraction of the cycle, so that the region over which N varies may have been too narrow to show up appreciably in the dHvA data. Possibly this puzzlement may reflect too naive an understanding of the QHE, but it would be of interest to see if the form of the dHvA oscillations approached that expected for ζ constant as the width of the QHE plateau approached the full width of a cycle. It would also be interesting to extend the dHvA study to the higher fields at which the fractional QHE appears, i.e. those such that F/B has the appropriate fractional values, e.g. 1/3; the published data do not go beyond about 9×10^4 G for which F/B ~1.3.

The interpretation of the intercalated graphite experiments is complicated by the eddy current effects characteristic of the modulation method and also perhaps by features of the domain boundary dynamics if domains due to MI do really occur. It would therefore be desirable to study the oscillations by a static technique such as the very sensitive torque methods developed by GRIESSEN [18] and by EISENSTEIN [19] for the experiments of [17]. Another interesting topic of study would be to see how the dHvA effect in conditions of strong MI changes as the sample thickness is reduced to become comparable with the cyclotron radius R. Perhaps such a study might be more effectively carried out in a 3-D metal such as silver for which strong MI is possible and R is of order 1 or 2 μ, rather than for an intercalated graphite for which R is of order 10 times smaller.

I should like to thank Dr I. D. Vagner and Dr R. S. Markiewicz for many stimulating discussions and the latter for sending me details of his work in advance of publication.

References

1. D. Shoenberg: Phil. Trans. Roy. Soc. A255, 85 (1962)
2. T.D. Holstein, R.E. Norton, P. Pincus: Phys. Rev. B8, 2649 (1973)
3. A.B. Pippard: Proc. Roy. Soc. A272, 192 (1963)
4. A.B. Pippard: In Electrons at the Fermi Surface, ed. by M. Springford (Cambridge University Press 1980) chap.4
5. D. Shoenberg: Magnetic Oscillations in Metals (Cambridge University Press 1984) chap.6
6. M.Ya. Azbel: Phys. Rev. B26, 3430 (1980)
7. I.D. Vagner, T. Maniv, E. Ehrenfreund: Phys. Rev. Lett. 51, 1700 (1985)
8. R.S. Markiewicz, M. Meskoob, C. Zahopoulos: Phys. Rev. Lett. 54, 1436 (1985)

9. R. Peierls: Z. Phys. 81, 186 (1933)
10. D. Shoenberg: J. Low Temp. Phys. 56, 417 (1984)
11. J.H. Condon: Phys. Rev. 145, 526 (1966)
12. J.H. Condon, R.E. Walstedt: Phys. Rev. Lett. 21, 612 (1968)
13. I. Privorotskii: Thermodynamic Theory of Domain Structures (John Wiley & Sons, New York, Toronto 1976)
14. I.D. Vagner: private communication
15. R.S. Markiewicz: preprint (1986) and private communication
16. F. Fang, P.J. Stiles: Phys. Rev. B28, 6992 (1983)
17. J.P. Eisenstein, H.L. Stormer, V. Narayanamurti, A.Y. Cho, A.C. Gossard, C.W. Tu: Phys. Rev. Lett. 55, 875 (1985)
18. R. Griessen: Cryogenics 13, 375 (1973)
19. J.P. Eisenstein: Appl. Phys. Lett. 46, 695 (1985)

REFLECTIONS ON HOLSTEIN'S ONE-DIMENSIONAL POLARON

Leonid A. Turkevich

In quasi-one-dimensional systems, the conditions for polaronic self-localization of electrons or holes via interaction with lattice vibrations is particularly favorable. The Holstein molecular crystal model provides an appropriate framework within which to study large polaron transport.

I want to start with a few personal remarks. Ted was extremely influential in my development as a physicist. His intellectual honesty set for me the highest standards--you knew you could not buffalo Ted. And woe to that seminar speaker who would try to gloss over a point that Ted had not understood--invariably the speaker had not either! Ted's tenacity bordered on stubbornness; like Jacob and the angel, he would not let go of a problem until he had wrestled it to the ground. This conveyed to his younger colleagues the confidence that no theoretical problem was ever too hard. However, Ted was also concerned with the question of "taste"; the talented should not work on just any problem. It was not unusual for Ted to walk into an office and solve in a morning a problem which had had one stymied for weeks; but those were <u>not</u> the problems worth working on. In the match between Jacob and the angel, it had better not be too easy for Jacob.

At UCLA, my office was next door to Ted's. Without fail, every morning Ted would make the rounds along the Solid State corridor to chat. No matter what initially might be on his mind, he would not leave until we had discussed physics. Even though our active collaboration only started during my last few months at UCLA, from the outset, Ted had kept me abreast of his 1d polaron work, and I would kibbitz on WKB solutions, a love for which we both shared.

Knowing nothing of polarons before coming to UCLA, I was frankly intimidated from getting involved in what was obviously a deep and subtle problem. For thirty years, Ted had been fascinated by the electron-phonon interaction and especially with its nonperturbative polaron consequences--almost to the point of obsession. Whenever a student would evince even a mild acquaintance with the polaron, Ted would zero in, "And what is your interest in the polaron?" By contrast, I remember one visitor opening a discussion in Ted's office with "What use is the polaron?" While the interaction remained civil, the visitor departed feeling like Pompeii after being worked over by Vesuvius. So if I were to do anything with Ted, it would have to be a polaron problem.

Quasi-one-dimensional solids [1] (anisotropic molecular crystals and especially polymeric crystals) are characterized by considerable anisotropy in their electronic transport properties. Within tight-binding theory, $J_{||}/J_{\perp} \gg 1$, where $J_{||}$ and $J_{\perp}$ are respectively the intrachain and interchain electron transfer amplitudes. Very general arguments [2-4] then indicate that, with any electron-phonon coupling, an excess electron or hole must occur in the form of a quasi-one-dimensional polaron, whose intrachain dimension L is large compared to the lattice spacing a, while still being confined to the single chain. The study of the dynamics of such anisotropic large polarons [5,6] necessitates the development of a formal theory [7] of the small oscillation behavior of the one-dimensional large polaron within a well-defined model, namely the Holstein molecular crystal model [8] with its one-dimensional large polaron solution.

The Hamiltonian for the molecular-crystal model [8] is given by

$$H = \sum_n \left[\frac{-\hbar^2}{2M} \frac{\partial^2}{\partial u_n^2} + \frac{1}{2} M\omega_o^2 u_n^2 \right] - J \sum_n a_n^+ (a_{n+1} + a_{n-1})$$

$$- A \sum_n u_n a_n^+ a_n \qquad (1)$$

The first term describes the vibrational motion of a chain of isolated diatomic molecules (mass M and Einstein frequency ω_o) as a function of the vibrational displacements, u_n, of the individual diatomic internuclear coordinates from their common equilibrium value. The second term describes electron-transfer between adjacent sites (as in conventional tight-binding theory) with overlap matrix element J; the notations a_n^+ and a_n denote fermion creation and annihiliation operators respectively. The third term constitutes the electron-lattice interaction; the Holstein molecular crystal model assumes this to be site-diagonal in the electron coordinate and purely local in that it depends only on the vibrational coordinate of the occupied site.

The zeroth order adiabatic approximation [9] consists of dropping the vibrational kinetic energy contribution to (1), namely,

$$T_L = \sum_n \frac{-\hbar^2}{2M} \frac{\partial^2}{\partial u_n^2} \quad . \tag{2}$$

The adiabatic one-electron wavefunction solves the Schrödinger equation

$$\left[\frac{1}{2} M\omega_o^2 \sum_p u_p^2 - Au_n\right] a_n(.u_m.) - J\left[a_{n-1}(.u_m.) + a_{n+1}(.u_m.)\right] = E(.u_m.) a_n(.u_m.) \tag{3}$$

Here, the one-electron wavefunction is specified in terms of the amplitudes, $a_n(..u_m..)$, for electron occupancy of the nth diatomic site; as in any adiabatic formulation, these amplitudes are parametric functions of the vibrational displacements, u_m. The eigenvalue, $E(..u_m..)$, also a parametric function of the displacements u_m, serves as the effective potential energy for vibrational motion. It is conveniently written as

$$E(..u_m..) = \frac{1}{2} M\omega_o^2 \sum_p u_p^2 + \epsilon(..u_m..) - 2J, \tag{4}$$

where $\epsilon(..u_m..)$ is the energy eigenvalue of the "electronic" equation, which, in the continuum approximation, appropriate for the case of the large polaron, takes the form

$$H(\{u_m\}) a_n(\{u_m\}) = -J \frac{\partial^2 a_n(\{u_m\})}{\partial n^2} - Au_n a_n(\{u_m\}) = \epsilon(\{u_m\}) a_n(\{u_m\}) , \tag{5}$$

wherein the aggregate, $\{u_m\}$, now corresponds to a continuous field variable. The electron-lattice interaction plays the role of an effective potential energy in the electron Hamiltonian.

The solution of (3) or of (5) for arbitrary displacements u_m is extremely difficult. However, within the Born-Oppenheimer approach [9], one first solves the problem for those values of u_m which minimize $E(\{u_m\})$, and then develops perturbative solutions for small displacements of the u_m about their equilibrium values $u_m^{(0)}$. The minimal energy solution was obtained by HOLSTEIN [7] and by RASHBA [10], which we now review.

The "equilibrium" displacements are given by

$$u_n^{(0)} = \frac{A}{M\omega_o^2} |a_n^{(0)}|^2 \quad , \tag{6}$$

Substituting (6) into (4), yields the nonlinear Schrödinger equation

$$J \frac{\partial^2 a_n^{(0)}}{\partial n^2} + \left[\frac{A^2}{M\omega_o^2} |a_n^{(0)}|^2 + \epsilon \right] a_n^{(0)} = 0. \tag{7}$$

In addition to the usual delocalized Bloch states, the nonlinear Schrödinger equation (7) possesses an exact self-trapped [11-13] solution,

$$a_n^{(0)} = (\tfrac{\gamma}{2})^{1/2} \operatorname{sech} [\gamma(n-\xi/a)], \tag{8}$$

The parameter $\gamma = A^2/4M\omega_o^2 J$ characterizes the size of the polaron and may be used to define a natural length scale transformation $z = \gamma n$. The variable ξ denotes the polaron-centroid coordinate; the translational invariance of the polaron is reflected in the ξ-independence of (7). Using (6), the equilibrium displacements are given by

$$u_n^{(0)} = \frac{A}{2M\omega_o^2} \gamma \operatorname{sech}^2[\gamma(n-\xi/a)] \tag{9}$$

The total energy (4) of the coupled electron-lattice system is

$$E_p = \epsilon + \frac{1}{2} M\omega_o^2 \sum_n |u_n^{(0)}|^2 = -J\gamma^2/3 \quad . \tag{10}$$

The energy E_p represents the polaron binding energy, relative to the corresponding minimum, -2J, for the delocalized Bloch electron state. The self-consistent electron-lattice interaction is

$$V^{(0)}(n) = -Au_n^{(0)} = -2\gamma^2 \ J \ \mathrm{sech}^2 \ [\gamma(n-\xi/a)]. \tag{11}$$

Using (11), one may rewrite (7) in the form

$$\frac{d^2a_n}{dn^2} + \left[2\gamma^2 \ \mathrm{sech}^2[\gamma(n-\xi/a)] - \gamma^2 \right] - a_n = 0. \tag{12}$$

We now consider the electronic problem for vibrational configurations in the vicinity of the minimal configuration, namely, for small departures $\delta u_m = u_m - u_m^{(0)}$ from the minimal configuration. We solve (12) perturbatively, obtaining corrections to the electron wavefunction a_n and energy ϵ to first and second orders in the δu_m, respectively. To second order in the small displacements δu_n, the total electron-lattice energy (4) is given by

$$E(..u_m..) = \epsilon^{(0)} + \epsilon^{(1)} + \epsilon^{(2)} - 2J + \frac{1}{2} M\omega_o^2 \sum_m \left[u_m^{(0)} + \delta u_m \right]^2 . \tag{13}$$

The cross-term in the last sum precisely cancels the first-order correction $\epsilon^{(1)}$ to the electronic energy, leaving

$$E(..u_m..) = -2J - J\gamma^2/3 + \frac{1}{2} M\omega_o{}^2 \left[\sum_n (\delta u_n)^2 - \sum_{nn'} g(n-\xi/a, n'-\xi/a)\delta u_n \delta u_{n'}\right] . \tag{14}$$

where

$$g(n,n') = 4\gamma^2 \ G(n,n') \ \mathrm{sech} \ \gamma n \ \mathrm{sech} \ \gamma n' \quad . \tag{15}$$

with the Green's function G(n,n') as that solution of the equation

$$\left[\frac{\partial^2}{\partial n^2} + 2\gamma^2 \mathrm{sech}[\gamma(n-\xi/a)] - \gamma^2\right] G(n,n') = -\delta(n-n') + \frac{\gamma}{2} \ \mathrm{sech}\gamma n \ \mathrm{sech}\gamma n' \tag{16}$$

which is orthogonal to the electronic ground-state wave function $a_n^{(0)}$.

We now investigate the normal modes of vibration associated with (14). The odd modes have previously been studied analytically by MELNIKOV [14], while only eigenfrequencies for the even modes have been treated numerically by SHAW and WHITFIELD [15]. The principal utility of these modes is their natural incorporation of translational invariance, thereby serving as a convenient vehicle for our study [6] of the interaction between the translational motion of the polaron and the other (non-translational) vibrational degrees of freedom. Assuming harmonic time dependence $e^{i\omega t}$ for the nuclear displacements, the small oscillations are the eigenmodes of the classical dynamical equations

$$M \frac{\partial^2}{\partial t^2} \delta u_n = - \frac{\partial V(..\delta u_m..)}{\partial u_n} . \tag{17}$$

Using (14) for the potential and rearranging

$$(\omega_o^2 - \omega^2)\delta u_n = \omega_o^2 \sum_{n'} g(n,n')\delta u_{n'} . \tag{18}$$

In writing down these expressions we tacitly consider the polaron centroid coordinate ξ as being fixed. Equation (18) has the form of a standard linear homogeneous integral equation, the solution of which, subject to a normalizability boundary condition, constitutes a complete set of normal-mode functions $u_\alpha(n)$, with associated eigenfrequencies ω_α. Defining

$$f(n) = \sum_{n'} G(n,n') \text{ sech } \gamma n' \, \delta u_{n'} , \tag{19}$$

the eigenmode is given by

$$\delta u_n = \frac{4\gamma^2}{1 - \omega^2/\omega_o^2} \text{ sech } \gamma n \;\; f(n) . \tag{20}$$

A direct consequence of the orthogonality condition is

$$\int_{-\infty}^{\infty} u_\alpha(n) \, dn = 0 . \tag{21}$$

Differentiating (19) and using (16) and (20), we have

$$\frac{d^2 f}{dz^2} + \left[(2 + \frac{4}{1-\omega^2/\omega_o^2}) \text{ sech}^2 z - 1 \right] f(z) = \frac{2 \text{ sech } z}{1-\omega^2/\omega_o^2} \int_{\infty}^{\infty} \text{sech}^3 z' \, f(z') dz' . \tag{22}$$

Equation (22) constitutes an integro-differential equation which, with appropriate boundary conditions (normalizability and the orthogonality condition) is equivalent to the integral equation (18). As (22) is invariant with respect to the replacements $z \to -z$, $z' \to -z'$, the eigenfunctions $f_\alpha(z)$ may be classified as either even or odd.

For the odd-parity solutions, as first obtained by MELNIKOV [14], the integral term on the right-hand-side of (22) vanishes. Introducing the transformation, $u = \tanh z$, yields

$$\frac{d}{du}\left[(1-u^2)\frac{df_\lambda}{du}\right] + \left[\lambda(\lambda+1) - \frac{1}{1-u^2}\right] f_\lambda(u) = 0 \quad . \tag{23}$$

The normalizable eigenfunctions of (23) with odd parity are the associated Legendre polynomials

$$f_\lambda(u) = P_\lambda^1(u) = (1-u^2)^{1/2}\frac{dP_\lambda}{du} \quad , \tag{24}$$

with eigenvalues $\lambda = 2,4,6,..$ These odd parity solution trivially satisfy (21). The eigenvalue equation then takes the form

$$\omega_s^{\,2}/\omega_o^{\,2} = 1 - \frac{4}{s^2+5s+4} \quad . \qquad s = \lambda-2 = 0,2,4,... \tag{25}$$

We note that the lowest eigenfrequency, $\omega_{s=o}$, vanishes; the other eigenfrequencies converge with increasing s towards an accumulation point at ω_o.

The most important of these odd parity modes corresponds to the lowest frequency, namely

$$u_o(z) \sim \left.\frac{\partial u^{(0)}(z-\xi/a)}{\partial \xi}\right|_{\xi=0} \quad . \tag{26}$$

If we displace the polaron centroid by an infinitesimal increment, $\delta\xi$, we find the change

$$u^{(0)}(z-\delta\xi/L) - u^{(0)}(z) \sim u_o(z)\ \delta\xi. \tag{27}$$

Thus, the existence of a nonvanishing vibrational amplitude associated with the zero-frequency mode, $u_o(z)$, corresponds just to a rigid displacement of the self-consistent polaronic distortion, $u^{(0)}(z)$. This result, of course, expresses the translational invariance of the self-consistent solution.

We now study [8] the integro-differential eigenvalue equation (22) for the even modes. Defining

$$g(u) = (1-u^2)^{1/2} f(u) \tag{28}$$

and differentiating, (22) becomes

$$\frac{d}{du}\left[(1-u^2)\frac{dg'}{du}\right] + \lambda(\lambda+1)\, g'(u) = 0 \quad , \tag{29}$$

i.e. we recover the Legendre equation, where $g'=dg/du$. Thus

$$g(u) = \alpha\left[P_{\lambda+1}(u) - P_{\lambda-1}(u)\right] + \beta\left[P_{\lambda+1}(-u) - P_{\lambda-1}(-u)\right] + \gamma \quad . \tag{30}$$

The constant γ is determined [8] to be

$$\gamma = (\alpha+\beta)\ \frac{\sin\lambda\pi}{\pi}\ \frac{2\lambda+1}{\lambda(\lambda+1)} \tag{31}$$

Imposing the orthogonality condition (21), yields the eigenvalue equation

$$\frac{2}{\pi}\left[\psi(\lambda+1) - \psi(1)\right] = \tan\frac{\lambda\pi}{2} \quad . \tag{32}$$

The eigenvalue condition (32) is also satisfied by the odd modes, for which λ is an even integer. An evaluation [8] of the transcendental eigenvalue equation (32) yields eigenvalues identical to those found by purely numerical means [15].

In the usual theory of lattice dynamics, the determination of the normal modes of vibration constitutes the essential problem. Once these modes are found, the usual harmonic oscillator theory takes over, yielding the standard picture of noninteracting phonons. However, in the present problem, having found one mode--the translational mode--for which the "stiffness" constant, $M\omega^2_{\alpha=0}$, vanishes, we now encounter a fundamental difficulty. The motion associated with such a mode is clearly not oscillatory, as the "displacement" amplitude, Q_0, may become indefinitely large; the small amplitude assumption for the excursions, δu_n, about a fixed minimal pattern $u^{(0)}(n-\xi/a)$, with a fixed centroid coordinate ξ is manifestly inadequate--the amplitude of any component which contains the translational mode must be expected to get arbitrarily large. Hence the harmonic form of the potential function in (14) cannot describe the vibrational dynamics of the system in which extended translational motion of the polaron is involved. It is thus necessary to adopt a more flexible zeroth order approach in which the centroid is no longer fixed but is permitted to assume arbitrary values; this separates the ω=0 translational (Goldstone) mode from the other (small oscillation) modes. The centroid coordinate, ξ, thus becomes an independent dynamical variable, which may be treated <u>to all orders</u> in its amplitude. An arbitrary displacement, u_n, is now written in the form

$$u_n = u^{(0)}(n-\xi/a) + \sum_{\alpha=1}^{\infty} Q_\alpha \; u_\alpha(n-\xi/a) \tag{33}$$

where the mode amplitudes Q_α remain small. Note that the mode summation excludes any contribution of the translational mode, $u_o(n-\xi/a)$, since the possibility of purely translational motion is already taken into account by permitting ξ to be an arbitrary dynamical variable. This elimination of the translational mode from the α-sum automatically precludes the over-counting which would otherwise ensue from the introduction of an extra lattice dynamical variable on top of the already complete set constituted by the u_n.

By virtue of the ξ-dependence of the mode amplitude functions, u_α(n-ξ/a), the translational motion is no longer decoupled from the other non-translational degrees of freedom. This coupling becomes explicit when we transform the vibrational kinetic energy (2) from the variables u_n to the new coordinates ξ and Q_α. The transformed vibrational wavefunction acquires the prefactor $\sqrt{J} = |1 + M/M_p \, \Sigma_\alpha \, S_\alpha Q_\alpha|^{1/2}$, from the Jacobian of the transformation; the coefficients S_α are given by

$$S_\alpha = \sum_n \frac{\partial u^{(0)}(n-\xi/a)}{\partial \xi} \frac{\partial u_\alpha(n-\xi/a)}{\partial \xi} \tag{34}$$

and where the polaron "effective mass" is $M_p = M \, \Sigma_n^{*} \, [\partial u^{(0)}(n-\xi/a)/\partial\xi]^2$.

The transformed Hamiltonian is $H = H_o' + H_1$, where

$$H_o' = \sum_\alpha \hbar\omega_\alpha \, (b_\alpha^+ b_\alpha + 1/2) - \hbar^2/2M_p \, (\partial/\partial\xi - \sum_{\alpha\alpha'} G_{\alpha\alpha'} \, b_\alpha^+ b_{\alpha'})^2 \tag{35}$$

and

$$H_1 = \frac{-\hbar^2}{8M_p}\left[\frac{1}{J}\left(\frac{\partial}{\partial\xi} - \sum_{\alpha\alpha'} G_{\alpha\alpha'}Q_{\alpha'}\frac{\partial}{\partial Q_\alpha}\right)+\left(\frac{\partial}{\partial\xi} - \sum_{\alpha\alpha'} G_{\alpha\alpha'}Q_{\alpha'}\frac{\partial}{\partial Q_\alpha}\right)\frac{1}{J}\right]^2$$

$$+ \frac{\hbar^2}{2M_p}\left(\frac{\partial}{\partial\xi} - \sum_{\alpha\alpha'} G_{\alpha\alpha'}Q_{\alpha'}\frac{\partial}{\partial Q_\alpha}\right)^2 - \frac{\hbar^2}{8M}\sum_\alpha (M/M_p)^2 S_\alpha^2 \left(\frac{1}{J} - 1\right)^2 \qquad (36)$$

where the matrix elements $G_{\alpha\alpha'}$ are defined by

$$G_{\alpha\alpha'} = \sum_n u_\alpha(n-\xi/a)\frac{\partial u_{\alpha'}(n-\xi/a)}{\partial\xi} \qquad (37)$$

and where b_α^+ and b_α are respectively boson creation and annihilation operators associated with the normal mode excitations,

$$b_\alpha = (M\omega_o/2\hbar)^{1/2} Q_\alpha + (\hbar/2M\omega_o)^{1/2}\,\partial/\partial Q_\alpha$$

$$b_\alpha^+ = (M\omega_o/2\hbar)^{1/2} Q_\alpha - (\hbar/2M\omega_o)^{1/2}\,\partial/\partial Q_\alpha \qquad (38)$$

For the adiabatic regime, the vibrational quantum $\hbar\omega_o$ is small compared to the polaron binding energy E_p, and H_1 may be treated as a perturbation relative to H_o'. To lowest order, $J \to 1$ and $H_1 \to 0$; expansion of J in powers of the operator $M/M_p\Sigma_\alpha S_\alpha Q_\alpha$ leads to a perturbation development in the small dimensionless ratio $(\hbar\omega_o/E_p)^{1/2}$ and gives rise to Cherenkov phonon emission and absorption processes.

It useful to rewrite $H_o' = H_o + H_2$ in terms of "local" creation and annihilation operators

$$b(\eta) = \sum_{\alpha=1}^{\infty} u_\alpha(\eta)\, b_\alpha \qquad b^+(\eta) = \sum_{\alpha=1}^{\infty} u_\alpha(\eta)\, b_\alpha^+ \qquad (39)$$

where we have introduced the "local-site" coordinate $\eta = n - \xi/a$, and where

$$H_o = \frac{-\hbar^2}{2M_p}\left[\frac{\partial}{\partial\xi} - \sum_\eta b^+(\eta)\frac{db(\eta)}{d\eta}\right]^2 + \sum_\eta \hbar\omega_o b^+(\eta)b(\eta)$$

$$- \frac{\hbar\omega_o}{2}\sum_{\eta\eta'} g(\eta,\eta')\, b^+(\eta)\, b(\eta') \qquad (40)$$

and

$$H_2 = -\hbar\omega_o/4 \, \Sigma \, g(\eta,\eta') \, [b^+(\eta)b^+(\eta') + b(\eta)b(\eta')] \qquad (41)$$

In order to clarify the physical content of the Hamiltonian (40-41), it is instructive to write down an expression for the centroid velocity operator

$$\dot{\xi} = i/\hbar \, [H_o,\xi] = -i\hbar/M_p \, [\partial/\partial\xi - 1/a \sum_\eta b^+(\eta)db(\eta)/d\eta] \qquad (42)$$

The first term exhibits the expected connection between velocity and the canonical momentum $p_\xi = -i\hbar \, \partial/\partial\xi$. We note that p_ξ is a constant of the motion, as is easily verified by evaluating its commutator with $H_o' + H_1$. This constancy arises, of course, from the invariance of the total Hamiltonian with respect to translation of both polaron and associated phonons (i.e. non-translational vibrational excitations). Remember that our description of the non-translational vibrational subsystem is formulated in terms of the normal modes, which are "attached" to the centroid coordinate, ξ. The difference between the kinetic momentum, $M_p\dot{\xi}$, and the canonical momentum, p_ξ, is essentially the momentum associated with the excitations of the non-translational (phonon) excitations:

$$P_{ph} = -i\hbar/ \, \Sigma_\eta \, b^+(\eta) \, db(\eta)/d\eta \qquad (43)$$

Were $b^+(\eta)$ and $b(\eta)$ true boson operators, (43) would obviously correspond to phonon momentum. However, since the sums in (39) go over an incomplete set of mode-functions (the translational mode-function being omitted), the $b(\eta)$ and $b^+(\eta)$ do not obey boson commutation rules:

$$[b(\eta), \, b^+(\eta')] = \delta_{\eta\eta'} - u_o(\eta)u_o(\eta') \qquad (44)$$

Of course, at distances far ($\gg 1/\gamma$) from the polaron centroid, the correction to the boson commutation rules may be ignored; for such phonon wave-packets, the interpretation of (43) as phonon momentum is acceptable.

From the above discussion, it is clear that the first term in (40), the polaron kinetic energy, is just the kinetic energy, $T = 1/2\ M_p\ \dot{\xi}^2$, associated with motion of the centroid coordinate, ξ. Its dependence (42) on the variables of the phonon field constitutes a major structural feature of the present theory. The term $\sum_\eta \hbar\omega_o\ b^+(\eta)\ b(\eta)$ expresses the energy of the "free" phonon field--i.e. that produced by vibrational excitations at large distances ($|\eta| \gg 1/\gamma$) from the polaron. As expected, these excitations are characterized by the frequency ω_o of the host crystal. The last term in (40) gives rise to intersite transitions of phonons (where both involved sites, η and η', lie in the immediate vicinity of the centroid coordinate); these intersite transitions consititute the principle mechanism for polaron-phonon scattering, and hence for transport relaxation [6].

Similarly, the terms in (41) produce two-phonon creation and annihilation transitions in the vicinity of the polaron. The occurrence of such terms is not particularly surprising; they are generally present in standard problems of lattice dynamics (e.g., phonon scattering by imperfections). In the present theory, they can give rise to the added complication of real phonon generation and annihilation events. We note that these double emission and absorption processes are qualitatively different from the Cherenkov processes described by H_1 in that they are of the same order as the phonon scattering processes described by H_o.

The main thrust of the collaboration with Ted was to develop a theory [6] for the intrachain transport of large one-dimensional polarons. We have reviewed the minimal energy adiabatic solution for the Holstein large polaron and have solved [8] analytically the classical equations of motion for the normal modes of the displacements about the minimal solution, both determining the eigenspectrum and constructing the eigenfunctions. Both eigenspectrum and eigenfunctions are essential for any systematic discussion of the transport and dynamics of such one-dimensional topological entities. A natural result of our small oscillation analysis is the appearance of the translational (zero frequency) Goldstone mode for the polaron. As this mode lacks a restoring force, it must be treated to all orders in its amplitude, in any consistent discussion of transport and dynamics. We have verified the basic quasi-free, band-type character of intrachain motion and have developed a theoretical framework for the treatment of the different contributions to transport relaxation. We have obtained [6] preliminary results on relaxation due to phonon-polaron scattering, Cherenkov phonon emission and absorption, and coherent double emission and absorption processes.

REFERENCES

1. For recent reviews see Molecular Crystals and Liquid Crystals 77 (1981); Physics in One Dimension, ed. by J. Bernasconi and T. Schneider (Springer Verlag, 1981); The Physics and Chemistry of Low Dimensional Solids, ed. by L. Alcacer (Reidel, 1980).

2. D. Emin & T. Holstein, Phys. Rev. Lett. 36, 323 (1976).

3. L.A. Turkevich & M.H. Cohen, unpublished.

4. B. Pertzsch & U. Rossler, Solid State Commun. 37, 931 (1981).

5. T. Holstein, Molecular Crystals & Liquid Crystals 77, 35 (1981).

6. T. Holstein & L.A. Turkevich, Physics Reports, to be published.

7. T. Holstein, Annals of Physics 8, 325 (1959).

8. L.A. Turkevich & T. Holstein, submitted to Physical Review B.

9. M. Born & J.R. Oppenheimer, Ann. Physik 87, 457 (1927).

10. E.I. Rashba, Optika i Spektroskopiya 2, 75 (1957).

11. L.D. Landau, Phys. Zeits. d. Sowjetunion 3, 664 (1933).

12. S.I. Pekar, Issledovaniya po elektronnoi teorii kristallov (Gostekhizdat, 1951), Ch. 2 (Untersuchungen über die Electronentheorie der Kristallie, Akademie-Verlag, Berlin 1954; Research in Electron Theory of Crystals, AEC Division of Technical Information, Washington D.C., 1963, transl. series no. AEC-tr-5575 Physics).

13. Y. Toyozawa, Progr. Theor. Phys. Japan 26, 9 (1961); and in Polarons and Excitons, ed. by C.G. Kuper & G.D. Whitfield (Oliver & Boyd, Edinburgh, 1963), p. 211.

14. V.I. Melnikov, Sov. Phys.--JETP 45, 1233 (1977) [Zh. Eksp. Teor. Fiz. 72, 2345 (1977)].

15. P.B. Shaw & G. Whitfield, Phys. Rev. B 17, 1495 (1978).

Spin-Spin Interactions and Two-Dimensional Ferromagnetism

Y. Yafet

It is widely known that a two dimensional lattice of spins coupled by Heisenberg (positive) exchange interactions has no long-range ferromagnetic order [1]. >From the point of view of the excitations, the quadratic dependence of the spin-wave frequency on the wave vector results in a divergence of the number of excited spin waves at any finite temperature T, thus precluding any long-range magnetization. Because the divergence is only logarithmic in the spin wave energy, it has been possible and customary to obtain a finite magnetization by introducing various cut-offs such as an anisotropy field or a finite sample size. The role of the spin-spin interactions, which are present in every magnetic material, does not seem to have been investigated even though they are anisotropic and might, as a consequence, remove the divergence of the spin deviation at finite T.

As they are two-body interactions, the spin-spin interactions cannot be treated as an effective field. They were first included in the dynamics of the spin system in the classic paper of HOLSTEIN and PRIMAKOFF [2]. Because the following calculation is a direct extension of the treatment of that paper, it seemed appropriate to include it here as a contribution to the Holstein symposium.

We follow the notation of [2] and refer the reader to that paper for the derivation of the expressions that we borrow from there. Consider an infinite square net of spins S with lattice constant a. Let $M_o = 2\beta S/a^2$ be the saturation moment per unit area, β being the Bohr magneton. Equation (23) of [2] gives the equilibrium value of the normalized spin deviation from saturation at temperature T:

$$\begin{aligned} \Delta m(T) &\equiv \frac{\Delta M(T)}{M_o} \qquad (1) \\ &= \frac{1}{(2\pi)^2 S} \int dK \, \frac{A(K)}{\sqrt{A(K)^2 - |B(K)|^2}} \, \frac{1}{\exp\left(\frac{\sqrt{A(K)^2 - |B(K)|^2}}{k_B T}\right) - 1} \end{aligned}$$

where the dimensionless quantity $\mathbf{K} = \mathbf{k}a$ and $\mathbf{k}$ is the wave vector in the plane xz of the net. The quantities A(K) and B(K) are given by:

$$A(\mathbf{K}) = 2\,S \sum_{\boldsymbol{\ell}} J(\boldsymbol{\ell})\,(1 - \exp\,(i\,\mathbf{K}\cdot\boldsymbol{\ell})) + 2\,\beta\,H$$
$$- \frac{4\,\beta^2\,S}{a_3} \sum_{\boldsymbol{\ell}} \frac{1}{\ell^3} \left(- \frac{3\ell_z^2}{\ell^2} \right) \tag{2a}$$
$$- \frac{2\,\beta^2\,S}{a^3} \sum_{\boldsymbol{\ell}} \frac{1}{\ell^3} \left(1 - \frac{3\ell_z^2}{\ell^2} \right) \exp\,(i\,\mathbf{K}\cdot\boldsymbol{\ell})$$

$$B(\mathbf{K}) = \frac{6\,\beta^2\,S}{a^3} \sum_{\boldsymbol{\ell}} \frac{\ell_x^2}{\ell^5} \exp\,(i\,\mathbf{K}\cdot\boldsymbol{\ell}) \tag{2b}$$

where $\boldsymbol{\ell}$ is a vector whose components ℓ_x, ℓ_y are integers, the point at the origin being excluded. The first line of (2a) includes the exchange energy which, for nearest neighbor interactions J reduces to $\alpha\,K^2$ with $\alpha = 2\,S\,J$, and the Zeeman energy which will be taken to be zero. The remaining terms of (2) come from the spin-spin interactions and they are anisotropic. The second line is the contribution of the dipolar field H_{dip} of the fully aligned spins (z is the direction of the magnetization) while the third line comes from the $S_{\ell}^{+}\,S_{m}^{-}$ terms and thus includes the dynamics of the spins. Likewise the term B(K) corresponds to the terms that flip two spins, $S_{\ell}^{-}\,S_{m}^{-}$ and $S_{\ell}^{+}\,S_{m}^{+}$. Replacing the dipolar interaction by an effective field would amount to neglecting the last two lines of (2) and would give a gap $2\,\beta\,H_{dip}$ in the spin wave spectrum. In contrast, as seen from (1) and (2), the dipolar interaction changes the energy of a spin wave K from A(K) to $\left[A(K)^2 - |B(K)|^2\right]^{1/2}$, a change which is important only for spin waves of small K because of the smallness of the dipolar interaction.

In three dimensions the term ℓ_x^2 in B(K) is replaced by $\ell_x^2 - \ell_y^2 - 2\,i\,\ell_x\,\ell_y$ and the dipolar sums become dependent on the shape of the sample. The calculation in [2] was carried out for an ellipsoid elongated in the direction of the magnetization. In the limit that the demagnetizing coefficient vanishes, A(K) equals $|\beta(\mathbf{K})|$ and there is no gap. For other sample shapes there will be a gap, but since $\Delta\,M(T)$ is not divergent in three dimensions the effect of a gap of that magnitude would be negligible except at temperatures of a fraction of a degree.

In two dimensions the situation is quite different: The lattice sums in (2) converge absolutely, there is no shape dependence, and for lattices with sufficient symmetry (square or triangular lattices) the equality

$$2 \sum_{\boldsymbol{\ell}} \left(\ell_x^2/\ell^5 \right) = 2 \sum_{\boldsymbol{\ell}} \left(\ell_z^2/\ell^5 \right) = \sum_{\boldsymbol{\ell}} 1/\ell^3$$

holds. This results in the equality $A(0) = B(0)$ which shows that the dipolar interaction produces no gap in two dimensional symmetric lattices. To see whether the divergence of $\Delta\,M(T)$ is removed, it is necessary to examine the K dependence of A(K) and B(K) for small K. Expanding the Bose factor in (1), we define

$$\Delta\,m(T)_{\ell} = \frac{k_B T}{(2\pi)^2 S} \int_0^{2\pi} d\phi \int_0^{q} \frac{A(\mathbf{K})K\,dK}{A(\mathbf{K})^2 - B(\mathbf{K})^2} \tag{3}$$

where the choice of the upper limit q will be discussed below. The dependence of the denominator on K will determine whether $\Delta\ m(T)_\ell$ converges. This dependence is calculated as follows: First, the lattice sums in (2) are calculated numerically for $K = 0$. The sum $\sum (1/\ell^3)$ is just a measure of the local field at a lattice site when the spins point along the normal to the layer. For a thin film (3-dimensional) this sum would be equal to the difference between the demagnetizing coefficient 4π and the coefficient of the Lorentz field, $4\pi/3$. For the square net the calculated sum is equal to $(8\pi/3)f$ where the numerical factor $f = 1.078$ is surprisingly close to unity. (For the triangular lattice this factor is 1.318, the larger value reflecting the close-packing of the lattice). Second, for $K \neq 0$, the sum over $\boldsymbol{\ell}$ is replaced by an integral over $\boldsymbol{\ell}$, which can be done analytically, with a spherical cut-off at r_o as the lower limit of ℓ. The value of r_o is chosen such as to reproduce the correct value of the sum at $K = 0$, i.e. $r_o = 3/(4f)$. The validity of this approximation for $K \neq 0$ has been checked numerically and found to be excellent. We thus obtain for the sum in (2b):

$$F(\mathbf{K}) = \sum \frac{\ell_x^2}{\ell^5} \exp\left(i\,\mathbf{K}\cdot\boldsymbol{\ell}\right) \simeq \int_{r_o}^{\infty} \frac{d\ell}{\ell^2} \int_0^{2\pi} \sin^2(\theta + \phi) \exp\left(i\,K\ell \cos\phi\right) d\phi \tag{4}$$

where ϕ is the angle between $\boldsymbol{\ell}$ to $\mathbf{K}$ and θ is the angle between $\mathbf{K}$ and the magnetization. Making use of the relation [3]

$$\exp\left(i\,x \cos\phi\right) = J_o(x) + 2\sum_{n=1}^{\infty} i^n J_n(x) \cos(n\,\phi)$$

we obtain

$$F(K) \simeq \pi \int_{r_o}^{\infty} \frac{d\ell}{\ell^2} \left[J_o(K\ell) + J_2(K\ell)\cos 2\theta\right]. \tag{5}$$

Further, integrating by parts and noting that $(dJ_o(x)/dx) = -J_1(x)$ and $\int_0^{\infty} (J_1(x)/x)\,dx = 1$, we finally obtain to first order in K

$$A(\mathbf{K}) = \alpha K^2 + 2\beta H + 4\pi\beta M_B f \left(1 - (K/4f)(1 + \cos 2\theta)\right), \tag{6a}$$

$$B(\mathbf{K}) = 4\pi\beta M_B f \left(1 - (K/4f)(3 - \cos 2\theta)\right), \tag{6b}$$

where $M_B = 2\beta S/\alpha^3$ is the magnetization. Letting $E_M = 4\pi\beta M_B$, we obtain for $H = 0$:

$$A(\mathbf{K})^2 - B(\mathbf{K})^2 = \alpha K^2 \left(\alpha K^2 + 2E_M\left(1 - (K/4f)(1 + \cos 2\theta)\right)\right)$$

$$+ E_M^2 (K/f)(1 - \cos 2\theta). \tag{7}$$

This expression has a term linear in K which is surprising at first glance since each $\boldsymbol{\ell}$ in (2a) and (2b) depends on $\mathbf{K}$ through $\cos(\mathbf{K}\cdot\boldsymbol{\ell})$ which contains only even powers of K. (The terms in $\sin(\mathbf{K}\cdot\boldsymbol{\ell})$ drop out of inversion symmetry.) This linear dependence is due to the long-range character of the dipolar interaction. It has been obtained by integrating (4) with a finite K in an infinite lattice. If the lattice is taken to be finite in size, say of radius La with $L >> 1$, then the linear dependence in (7) holds for $KL >> 1$. In the opposite case $KL << 1$, the dependence on K is K^2L. It is easy to see that the contribution to (1) from the K values of order $n\pi/L$ with $n \lesssim 10$ is of order $(1/L)$ of the contribution from all the other K values which satisfy $KL >> 1$. Hence doing the calculation with an infinitely extended lattice gives the correct answer. Thus the long-range character of the dipolar interaction is what is ultimately responsible for the modification of the spin wave spectrum near $K = 0$.

In the limit of small K, the integrand reduces to $[2\,\alpha K^2 + (E_M/f)\,K\,(1 - \cos 2\theta)]^{-1}$, and doing the angular integration first we find

$$\Delta\, m(T)_\ell = \frac{k_B T}{2\,\pi\, S\, \alpha}\, \ell n \left[(1 + \alpha q f/E_M)^{1/2} + (\alpha q f/E_M)^{1/2}\right], \tag{8}$$

which for $\alpha q/E_M << 1$ reduces to

$$\Delta\, m(T)_\ell \simeq \frac{1}{2\,\pi\, S}\, \frac{k_B T}{\sqrt{4\pi\beta M_B \alpha}}\, \sqrt{q}\ . \tag{9}$$

The convergence of the integral shows that in the linearized spin wave approximation, the dipolar interaction leads to ferromagnetism in a two-dimensional lattice of spins with short-range Heisenberg interactions. For K values larger than q the evaluation of (1) must be done numerically. It is possible to choose q small enough that the expansion of the Bose factor is valid, and at the same time large enough for the variation of $A(K)^2 - |B(K)|^2$ with the angle θ to be negligible. For such a q the angular integration gives a factor 2π and, for $\mathbf{K} > q$, (1) reduces to a simple integral over K which has been done numerically. The results are shown in Figures 1 and 2 for spins $S = 1/2$ and $S = 7/2$ respectively. Also shown there are calculations for films of 2, 3, and 4 atomic layers. Details on these calculations can be found elsewhere [4].

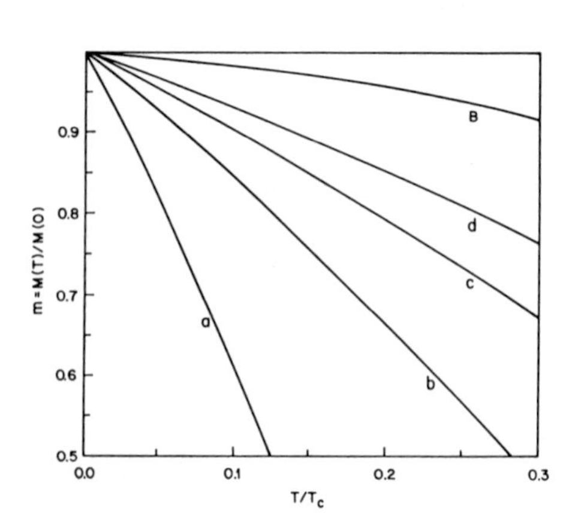

Figure 1. Normalized magnetization versus reduced temperature for $S = 1/2$. Curves a, b, c, d are for 1, 2, 3, 4 square atomic layers respectively. B is the bulk $T^{3/2}$ law.

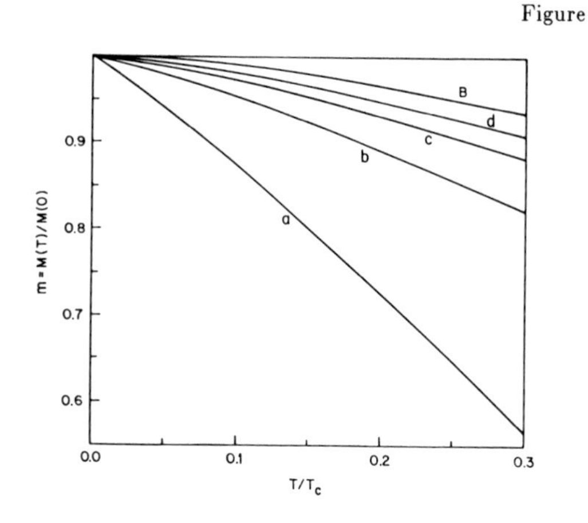

Figure 2. Normalized magnetization versus reduced temperature for S = 7/2. Curves a, b, c, d are for 1, 2, 3, 4 square atomic layers respectively. B is the bulk $T^{3/2}$ law corrected for the influence of the dipolar interaction.

It is seen from Figs. 1 and 2 that the magnetization falls off with T much more slowly for S = 7/2 than it does for S = 1/2. The reason is that the ratio of dipolar to exchange energy is much larger for the larger spin: Thus, for fixed Curie temperature, J is essentially proportional to $[S(S+1)]^{-1}$ and hence $\alpha = 2\,S\,J$ is inversely proportional to S + 1 while E_M is proportional to S. The ratio E_M/α is then $\simeq$ 20 times larger for S = 7/2.

The present work was motivated by measurements of the magnetization M(T) in metallic superlattices of ferromagnetic Gd (S = 7/2) and nonmagnetic Y. I am indebted to Dr.'s J. Kwo and E. M. Gyorgy for drawing my attention to this problem.

1. F. Keffer in Handbuch der Physik, vol. XVIII/2, edited by S. Flügge (Springer-Verlag, New York, 1966).
2. T. Holstein and H. Primakoff. Phys. Rev. 58, 1098 (1940).
3. F. Bowman, Introduction to Bessel Functions (Dover, New York, 1958).
4. Y. Yafet, J. Kwo, and E. M. Gyorgy, Phys. Rev. B33, 6519 (1986).

Fractons and the Ioffe-Regel Limit

O. Entin-Wohlman

Abstract

The fracton description of vibrational excitations in amorphous systems is reviewed. In particular, the lifetime of the vibrational modes, required for the interpretation of the line width in scattering experiments, is analyzed, with emphasis upon the strong scattering regime. A new argument suggests that fractons are always in the strong scattering, Ioffe-Regel limit. Their relevant fracton dimensionality then has a universal value of 4/3. Comparison is made with experimental data.

1. Introduction

The problem we want to discuss is that of localized vibrational excitations which are strongly perturbed by structural inhomogeneities.

Localization in disordered systems can be described in terms of characteristic length scales. The extended phonons of an ordered solid have wave lengths λ_{ph}, inversely proportional to their frequency ω. When they are slightly perturbed, they acquire a finite mean free path ℓ, such that $\ell \gg \lambda_{ph}$. When the disorder is increased and reaches a critical value, multiple scattering Anderson localization sets in and phonons of frequencies above the mobility edge frequency (determined by the disorder) become localized. This introduces a third length scale – the localization length λ_L. At the vicinity of the Anderson mobility edge the three length scales are different, satisfying the inequality $\lambda_{ph} < \ell < \lambda_L$. They all decrease as the frequency is further increased beyond the mobility edge. The mean free path decreases faster than λ_{ph}, becoming comparable to it when the inverse lifetime of the vibrations, $\tau^{-1}(\omega)$, is of the order of ω. This is the condition which defines the phonon Ioffe-Regel frequency, ω_{IR}. At frequencies higher than ω_{IR}, the above weak scattering theory description loses its meaning. One expects that, at about ω_{IR}, all three length scales become comparable, and the vibrational modes are described by a single length scale. They are, of course, localized, but the nature of localization is very different from Anderson localization, being instead a consequence of strong scattering. We propose that localized vibrations characterized by a single length scale can be consistently described by the scaling model for fractons.

The Ioffe-Regel frequency ω_{IR} separates the vibrational spectrum into two regimes: weak scattering ($\omega < \omega_{IR}$), and strong scattering ($\omega > \omega_{IR}$). Anderson localization occurs in the weak scattering limit, where the three length scales discussed above have separate identities. Fractons prevail in the strong scattering regime where there is only a single length scale. This difference has many implications. In this article we concentrate upon the behavior of the vibrational lifetime, relevant to the measured line width in various scattering experiments (such as neutron or Brillouin scattering). We shall argue that the fracton lifetime, in the context of the Ioffe-Regel condition, implies a universal value for the fracton dimensionality, $\bar{\bar{d}}$.

One of the consequences of the weak scattering nature of Anderson localization is that the localized phonon density of states maintains its scattering-free form, $N(\omega) \sim \omega^{d-1}$, where d is the Euclidean dimensionality. The fracton density of states is characterized by the fracton dimensionality $\bar{\bar{d}}$, which in random systems is different from d. A scaling argument for the fracton lifetime to be presented below will require $\bar{\bar{d}} = 4/3$.

The specific numerical value of $\bar{\bar{d}}$ has attracted considerable interest in recent years because, for percolation clusters generated in spaces of dimension $d \geq 2$, it appears to have a universal value very close (exactly equal?) to 4/3. Our calculation below shows that this value has a direct relationship to the Ioffe-Regel condition for strong scattering.

In Section 2 we describe the Ioffe-Regel limit for vibrational modes, and present experimental indications of its existence in amorphous materials. In Section 3 we review the prominent characteristics of the scaling model for the fracton vibrational spectrum. The lifetime of the vibrations is treated in Section 4. We describe recent experimental data supporting our predictions and our conclusions in

Section 5.

2. The Ioffe-Regel Condition for Strong Scattering

In the ordinary description of plane wave-like excitations, a small perturbation on the system results in a finite width $\hbar/\tau$ of the energy levels, such that for a level of energy $E = \hbar\omega$

$$\omega > 1/\tau(\omega) \quad . \tag{1}$$

Dividing this inequality by the excitation velocity c one finds

$$\ell > \lambda \quad , \tag{2}$$

where $\ell = c\tau$ is the mean free path and $\lambda = c/\omega$ is the wave length. The Ioffe-Regel criterion [1], well known in the context of electronic transport, states that the weak scattering description ceases to be valid once λ becomes comparable to ℓ. For phonons, the Ioffe-Regel condition becomes

$$\omega\tau(\omega) \sim 1 \quad . \tag{3}$$

This value for ω marks the beginning of the strong scattering frequency regime. This regime hardly exists for metallic electrons. Their relevant length is the Fermi wave length, of the order of the lattice spacing. The Ioffe-Regel limit is thus reached near the zone boundary.

The situation may be quite different for vibrational excitations in amorphous materials. Analysis of experimental data indicates that a strong scattering frequency regime over a large part of the vibrational frequency space exists for glasses and other amorphous materials. In particular, estimates of the phonon mean free path for a variety of glasses have been deduced from thermal conductivity data [2,3] using the kinetic heat-conduction formula. They suggest that phonons of frequency roughly equal to the "plateau" temperature possess mean free paths comparable to their wave lengths. This occurs for λ of the order 20-50 Å. This value is larger than the structural unit length of these materials, of about 4 Å, by roughly an order of magnitude. Thus vibrations in glasses may exhibit the strong scattering frequency regime over a large fraction of the vibrational excitation spectrum.

A weak structural disorder imposed upon the low frequency vibrations leads to Rayleigh elastic scattering [4]

$$\frac{1}{\tau} = \omega\left(\frac{\omega}{\omega_{IR}}\right)^{d} \quad , \tag{4}$$

where d denotes the Euclidean dimensionality. When at $\omega = \omega_{IR}$ the corresponding wave length is still larger than the microscopic structural length, there is a regime, ranging from ω_{IR} up to the upper cutoff frequency (the Debye frequency) in which the scattering is too strong to be described by the Rayleigh law. One is in the strong scattering, or Ioffe-Regel regime. This is portrayed in Fig. 1.

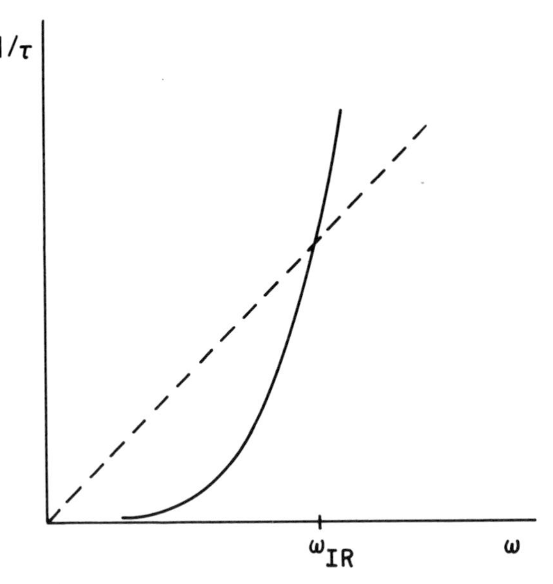

Fig. 1. Schematic plot of Rayleigh law [Eq. (4)] and the frequency (the dashed line).

There are virtually no microscopic theoretical models to treat this strong scattering regime. The detailed formulations of phonon localization [5-7] treat only Anderson localization. They are confined to the weak scattering regime of frequencies below ω_{IR}. These treatments implicitly assume that ω_{IR} is comparable to the Debye frequency (i.e., that there is no strong scattering regime) and introduce various expansions in a small disorder parameter [5,6]. As such, they cannot extend into the Ioffe-Regel regime.

One is thus faced with the problem of describing localized vibrational excitations above the Ioffe-Regel frequency where, presumably, they are characterized by a single length scale.

This type of vibration was first studied by Alexander and Orbach [8] in the context of the dynamics of self-similar fractal objects. For this reason, they gave these vibrations the name "fractons". The nature of the fracton spectrum was derived from scaling arguments, which so far have no detailed microscopic basis. However, the scaling model has a significant predictive power. One can calculate various properties from it which can be measured in random systems: the vibrational density of states and the temperature dependence of the specific heat [9,10], relaxation rate distributions [11], inelastic scattering of electrons off fractons [12], and the high-temperature thermal conductivity [13]. These can be compared with experimental data when available. The scaling model results, which in most cases are far from being obvious, appear to be in rather good agreement with experiment. Of particular interest in this respect is the high-temperature thermal conductivity K. The fracton model predicts a linear temperature dependence of K in the temperature range above the "plateau" due to phonon-assisted hopping of fractons. This appears to be confirmed by experiments [14] over a very wide temperature range.

The success of the fracton scaling model in comparing with experimental data taken on various random systems cannot, of course, be regarded as a conclusive proof for its validity. That is, it may not prove to be a unique description. However, one is encouraged to explore its implications further. Recently [15], the scaling model has been used to study the fracton lifetime. The results reveal an internal consistency of the model in a rather surprising manner, and provide an additional feature to be compared with experiment. We shall describe in the following Section the main properties of the fracton scaling model. We then will discuss the vibrational lifetime in the weak scattering regime, and in the Ioffe-Regel regime.

3. The Fracton Model

The scaling model for fractons is based upon the observation that many random systems, such as percolation clusters, polymers, rubbers, and gels, exhibit a crossover behavior. They appear to be homogeneous for length scales L large compared with some "correlation length" ξ, but have anomalous features for length scales L less than ξ. This is particularly manifested in vibrational excitations describing the elastic properties of the disordered medium [8,16]. The vibrations crossover from phonons, with long wave lengths (larger than ξ) to a new type of localized mode, fractons, with localization lengths smaller than ξ. In the vibrational problem, length scales can be translated into frequencies. Thus the correlation length is related to a crossover frequency ω_c, which separates the low frequency phonons from the (relatively) high frequency fractons:

$$\omega_c \sim \xi^{-(1 + \frac{\theta}{2})} \quad . \tag{5}$$

Here θ is the exponent characterizing the spatial decay of the classical diffusion coefficient [17], or, equivalently, the scaling properties of the elastic force constants [18].

The dispersion law at low frequencies is linear, like that of ordinary phonons,

$$\lambda_{ph} = c/\omega \sim \xi^{-\theta/2}/\omega \ , \qquad \omega < \omega_c \ , \tag{6}$$

but with a velocity of sound c which depends upon ξ. A scale-dependent velocity of sound has been recently measured by Courtens et al. [19] using Brillouin scattering from silica aerogels. These materials are porous, and are self-similar at length scales between the grain size and a certain cutoff length, determined by the gel density. We can associate ξ with the cutoff length. As the gel density decreases, ξ becomes larger. It was observed [19] that, at the same time, the velocity of sound decreases according to the prediction of Eq. (6).

The phonon regime, $\omega < \omega_c$, is thus characterized [see Eq. (6)] by two lengths; the wave length λ_{ph} and the correlation length ξ. The scaling model for fractons states that the excitations in the fracton regime, $\omega > \omega_c$, are characterized by a single length, the fracton localization length λ_{fr}. Assuming scaling, one accordingly replaces ξ and λ_{ph} in Eq. (6) by λ_{fr}, and obtains the fracton dispersion law

$$\lambda_{fr}^{-(1 + \frac{\theta}{2})} \sim \omega \ , \qquad \omega > \omega_c \ . \tag{7}$$

This shows a rather strong dependence of the localization length upon the frequency. It is this feature, combined with the assumption of a single length scale for fracton excitations, which distinguishes fractons from localized phonons treated in the Anderson sense [5]. The localization length of the latter, not too close to the mobility edge, is assumed to be almost frequency independent. The difference in the frequency dependences of the localization length is crucial. We have found different results for the Raman relaxation rate distributions between fractons and phonons localized in the Anderson sense [11]. The linear temperature dependence of the high-temperature thermal conductivity found for fractons [13] is not obtained for Anderson localized phonons.

The two dispersion laws, Eqs. (6) and (7), are mirrored in the form taken by the vibrational density of states. In the fracton regime, the density of states

$$N(\omega) \sim \omega^{\bar{\bar{d}}-1} , \qquad \omega > \omega_c , \tag{8}$$

is characterized by the novel dimensionality $\bar{\bar{d}}$. This form of the fracton density of states was calculated by Alexander and Orbach [8], who called $\bar{\bar{d}}$ the fracton, or spectral, dimensionality:

$$\bar{\bar{d}} = 2D/(2+\theta) , \tag{9}$$

where D is the fractal (Hausdorff) dimension. For many purposes, $\bar{\bar{d}}$ is the intrinsic relevant dimension. Generalization [20] of the Thouless argument for localization on a fractal object shows that $\bar{\bar{d}}$ replaces the Euclidean dimensionality, d, and should be less than 2 for localization, independent of the fractal dimension D. One notes that in Anderson localization for lattice vibrations in less than two dimensions, the mobility edge is at zero frequency, and the localization length diverges as the frequency is reduced [5]. This is analogous to the behavior depicted by Eq. (7), though of course the detailed functional form is not the same.

In the phonon regime the density of states has the ordinary d-1 power dependence upon the frequency. However, similar to the phonon dispersion relation (6), there is a prefactor depending upon ξ or ω_c [Eq. (5)]:

$$N(\omega) \sim \omega_c^{\bar{\bar{d}}-d} \, \omega^{d-1} , \qquad \omega < \omega_c . \tag{10}$$

The two forms for the density of states, Eqs. (8) and (10), are based upon a scaling argument, and consequently the crossover region around ω_c cannot be accurately described. However, the density of states should be normalized such that its integral up to the upper (microscopic) frequency cutoff will give the total number of modes per atom. Taking this into account, one obtains a rapid increase in the density of states at crossover [10], required for the interpretation of specific heat data, as measured in nearly all amorphous materials (for example, epoxy resins [24]).

Having set down the basic features of the fracton scaling model, we now proceed to investigate how scattering can be incorporated into the model, and the implications of the frequency dependence of the excitation lifetimes, particularly in the strong scattering regime.

4. Scattering in the Fracton Model

The vibrational spectrum is frequently probed by scattering experiments, e.g., neutron scattering which measures the imaginary part of the response function. In such experiments, the vibrations appear to have line widths which depend upon their frequencies. The line width is related to the lifetime of the excitations. To calculate the latter, we assume the existence of some scattering perturbation which may result from local fluctuations in the structure (e.g., in the atomic mass distribution or in the elastic force constants). This is analogous to the computation of the lifetime of ordinary phonons. Instead, here we want to evaluate this lifetime within the fracton model.

The crossover from phonons to fractons is based upon a scaling argument which assumes a single frequency ω_c separating the phonons from the fractons. In this spirit, the simplest way to consider the vibrational lifetime is to write down a scaling form [15],

$$\frac{1}{\tau} = \omega f\left(\frac{\omega}{\omega_c}\right) . \tag{11}$$

Then, Rayleigh scattering law [Eq. (4)], pertaining to the phonons, implies that $f(x) \sim x^d$ for $x \ll 1$. In the fracton regime, according to the scaling idea, there should be no dependence upon ω_c, and therefore $f(x) \to$ constant for $x \gg 1$. Thus, the scaling model yields

$$\frac{1}{\tau} \sim \omega_c^{-d}\, \omega^{d+1} , \qquad \omega < \omega_c , \tag{12a}$$

$$\frac{1}{\tau} \sim \omega \qquad , \qquad \omega > \omega_c . \tag{12b}$$

The scaling behavior implied by Eqs. (12) is depicted in Fig. 2. It has two strong implications. Equation (12a) combined with Eq. (4) implies that the Ioffe-Regel frequency of the phonons, ω_{IR}, is proportional to the crossover ω_c. Equation (12b) gives that fractons are always in the Ioffe-Regel limit, for all fracton frequencies ($\omega > \omega_c$). If these two conclusions did not hold, then scaling, at least for the scattering lifetime, would break down in the sense that a single characteristic frequency (ω_c), or alternatively a single correlation length (ξ), would not be sufficient to describe the vibrational modes.

There is some experimental evidence for the two conclusions drawn from Eqs. (12). The Brillouin scattering data of silica aerogels [19] yields the dispersion law and the inverse lifetime (the line width) for wave vectors smaller than ξ^{-1}, namely for the phonon regime. The dispersion law was observed to be linear in ω, and the line width

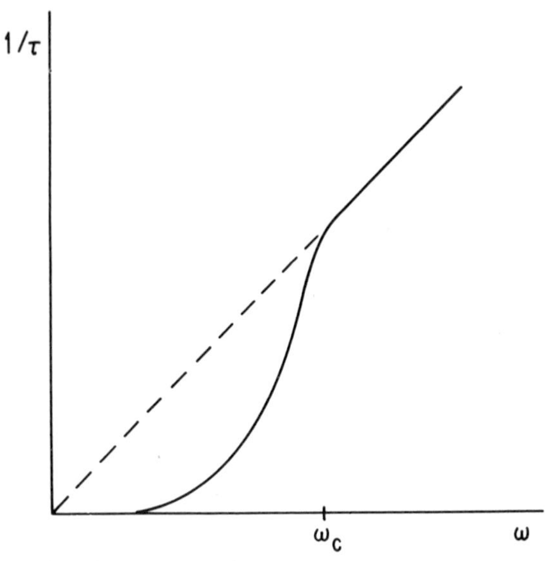

Fig. 2. Schematic plot of the vibration lifetime according to Eqs. (12). The dashed line is the frequency ω.

obeyed the Rayleigh law. Both were found to scale with respect to density, which in this material determines ξ and ω_c. Thus, a consistent interpretation of this experiment shows that, in the silica aerogels, the Ioffe-Regel frequency of the phonons is proportional to the crossover frequency. We shall come back to this point later. Unfortunately, the fracton regime of the silica aerogels has not yet been probed. However, the prediction that fractons are always in the Ioffe-Regel limit seems to be qualitatively consistent with recent neutron scattering results [22] on the dilute isotropic antiferromagnet $Mn_{0.5}Zn_{0.5}F_2$. The data yield the imaginary part of the response function as a function of the neutron frequency, for various values of the neutron wave vector. Magnons of a random antiferromagnet behave in many respects like the vibrational excitations of a random system [16]. Qualitatively, the neutron data exhibit the two main features discussed above: a crossover from phonon-like behavior at small wave vectors into a much different, localized-like behavior at larger neutron momentum transfer. The response function peak in the latter regime is very broad, with a line width of the same order of magnitude as the peak frequency itself.

Though these experimental results agree qualitatively with the predictions of scaling, Eqs. (12), one nevertheless feels that the latter should be carefully examined. One approach would be to derive the lifetime by some other method. We carry out such a calculation using the Golden Rule of time dependent perturbation theory:

$$\frac{1}{\tau} \sim |V|^2 \; N(\omega) \quad . \tag{13}$$

Here V represents a matrix element for the transition out of the initial state into a final state at the same frequency ω. The structural irregularities which give rise to the scattering potential V lead to the appearance of local strains, proportional to

$$\phi_i - \phi_j \sim \vec{a}_{ij} \cdot \vec{\nabla}\phi_i \quad , \tag{14}$$

where ϕ_i is the displacement of atom i, and $\vec{a}_{ij}$ is an appropriate microscopic unit vector. The matrix element V is proportional to the square of the local strain.

In the phonon regime the local strain scales as $\omega^{1/2}$. [This follows from the normal mode expansion of the displacement ϕ, where the amplitude of each normal mode scales like $\omega^{-1/2}$, and from the spatial derivative, related to λ_{ph}^{-1}, proportional to ω.] Consequently, the matrix element V scales like ω, as for ordinary phonons. We average the square of the matrix element over a region of linear size ξ and multiply by the phonon density of states [Eq. (10)], to obtain [15]

$$\frac{1}{\tau} \sim \omega^2 \; \xi^D \; \omega_c^{\bar{\bar{d}}-d} \; \omega^{d-1} \sim \omega_c^{-d} \; \omega^{d+1} \quad , \qquad \omega < \omega_c \; , \tag{15}$$

where we have used Eqs. (5) and (9). This reasoning reproduces the scaling form result, Eq. (12a) [23], and thus confirms the low frequency part of the scaling argument.

In the fracton regime, where scattering is presumably strong, one has to be cautious when using the perturbation theory expression. Our philosophy is to try to describe strong scattering in real (Euclidean) space by relatively weak scattering on a fractal geometry, and calculate the lifetime from the Golden Rule in the fractal geometry. Coarse-grain the fractal into units of linear size λ_{fr}. This requires the definition of an effective strain on the scale of the fracton localization length λ_{fr}.

Let us denote by R(x) an effective distance on the fractal correspond-ng to a distance x in Euclidean space. It can be shown [24] that R(x) is the distance along the bonds, and therefore proportional to the resistance between two points on the fractal at a Pythagorean distance x, when there is a unique connecting path. Thus, for example, the Heisenberg spin-spin correlation function, decays at low temperatures as a simple exponential of R(x), whereas in an ordered space it decays as a simple exponential of x [24]. In general, R(x) scales with x like the point-to-point resistance (or force constant in scalar elasticity [18]), required for the coarse-graining of the fractal into units of size x:

$$R(x) \sim x^{\bar{\zeta}}, \qquad \bar{\zeta} = (2-\bar{\bar{d}})D/\bar{\bar{d}} \quad . \tag{16}$$

We next calculate the matrix element V for a coarse-grained unit of linear size λ_{fr}. Since the relevant basic length scales like the resistance, one has

$$R(\lambda_{fr}) \sim \lambda_{fr}^{\bar{\xi}} \sim \omega^{2-\bar{\bar{d}}} \, , \tag{17}$$

using the fracton dispersion law [Eq. (7)] and Eqs. (9) and (16). This determines an effective frequency Ω, related to the basic length by a linear "dispersion law", $R(\lambda_{fr}) \sim \Omega^{-1}$, and hence related to the "real" frequency ω by

$$\Omega(\omega) \sim \omega^{2-\bar{\bar{d}}} \quad . \tag{18}$$

The new basic scales, $R(\lambda_{fr})$ and $\Omega(\omega)$, set the scale for the normal modes on the fractal. Therefore, the effective local strain on the scale of λ_{fr} will be proportional to $\omega^{-1/2}\,\Omega$, where $\omega^{-1/2}$ again results from the normal mode expansion, and Ω comes from the spatial derivative taken on the fractal. The latter replaces the ω-factor which appears in the analogous calculation for the phonons (and ordinary phonons as well). It follows that the matrix element V scales on the fractal as $\Omega^2/\omega \sim \omega^{3-2\bar{\bar{d}}}$, by Eq. (18). Squaring the matrix element and multiplying by the fracton density of states, Eq. (8), one obtains [15]

$$\frac{1}{\tau} \sim \omega^{5-3\bar{\bar{d}}} \, , \qquad \omega > \omega_c \; . \tag{19}$$

This agrees with the scaling argument result, Eq. (12b), and obeys the Ioffe-Regel condition, Eq. (3), only for $\bar{\bar{d}} = 4/3$.

In the phonon regime, $\omega < \omega_c$, the scaling argument gives the same result for the lifetime as a direct calculation of the Golden Rule. In the fracton regime, $\omega > \omega_c$, the scaling argument yields the Ioffe-Regel condition, Eq. (3). This can be combined consistently with the Golden Rule derivation of the fracton lifetime only when one forces in the latter the effective "quantum" fracton dimensionality $\bar{\bar{d}}$ of 4/3.

The numerical value of $\bar{\bar{d}}$ has been a subject of considerable controversy [25]. When they first introduce the fracton dimensionality, Alexander and Orbach [8] computed $\bar{\bar{d}}$ from Eq. (9) and noted that it was approximately 4/3 for percolating networks in all dimensions $d \geq 2$. This observation led them to suggest that $\bar{\bar{d}} = 4/3$ may be an exact relation (this is now known as the Alexander-Orbach conjecture). Of the many attempts to prove or disprove this numerical value [25], some have shown that $\bar{\bar{d}}$ is somewhat less than 4/3 in high dimensions, $d = 6-\varepsilon$, while others remain inconclusive. The numerical value of $\bar{\bar{d}}$ appears here rather surprisingly in the context of the fracton lifetime. Our calculation points out a strong connection between the general Ioffe-Regel criterion [Eq. (3)] and the 4/3 conjecture.

Values of $\bar{\bar{d}}$ different from 4/3 imply the breakdown of scaling, in the sense that a single characteristic frequency is not sufficient to describe the vibrational spectrum. One is forced to introduce new crossover frequencies. Such values of $\bar{\bar{d}}$ have still another complication. Let us write down the fracton lifetime, Eq. (19), in the form

$$\frac{1}{\tau} = \omega\left(\frac{\omega}{\omega_{IR}^{fr}}\right)^{4-3\bar{\bar{d}}} \tag{20}$$

so that the Ioffe-Regel limit is reached in the fracton regime only at a single frequency, ω_{IR}^{fr}. Then, for $\bar{\bar{d}} > 4/3$, the inverse lifetime $1/\tau$ exceeds the frequency ω for frequencies below ω_{IR}^{fr}. In this regime the scattering becomes too strong even on the fractal, and in complete analogy with the breakdown of Rayleigh law [Eq. (4)] above ω_{IR}, the scattering description becomes meaningless. The same argument holds for values of $\bar{\bar{d}}$ less than 4/3, at frequencies above ω_{IR}^{fr}.

One is therefore forced to assume that the Ioffe-Regel frequency of the phonons, ω_{IR}, is not proportional to the crossover frequency ω_c. Let us consider in detail the case of $\bar{\bar{d}} < 4/3$ (because whenever the Alexander-Orbach conjecture appears to break down, the fracton dimensionality is less than 4/3 [18,25]). Then, starting from small frequencies, the lifetime of the phonons is given by Rayleigh law, Eq. (4), in the frequency regime from 0 up to ω_c. At $\omega = \omega_c$ the lifetime crosses over to that of fractons whose $\bar{\bar{d}} < 4/3$, given by Eq. (20). Equating the two lifetimes at $\omega = \omega_c$ we obtain

$$\omega_{IR}^{d} \sim (\omega_{IR}^{fr})^{4-3\bar{\bar{d}}}\ \omega_c^{d-4+3\bar{\bar{d}}}\ , \tag{21}$$

which gives the scaling of the phonon lifetime with ω_c [see Eq. (4)]. It also shows that when $\bar{\bar{d}} < 4/3$, $\tau\omega_c$ does not scale as a function of ω/ω_c. (Note that ω_{IR}^{fr}, pertaining to the fractons, is not related to any macroscopic scale of the system, e.g., the correlation length ξ. In other words, it cannot be made critical.) In the frequency regime between ω_c and ω_{IR}^{fr} the lifetime is given by Eq. (20). At ω_{IR}^{fr}, the lifetime must crossover again to the Ioffe-Regel limit, $1/\tau \sim \omega$. This is because Eq. (20), with $\bar{\bar{d}} < 4/3$, is no longer valid as it contradicts the uncertainty principle ($\omega\tau \geq 1$). One is in the strong scattering regime and the value of $\bar{\bar{d}}$ must crossover to 4/3. We depict the behavior of $1/\tau$, implied from Eqs. (4), (20), and (21) in Fig. 3.

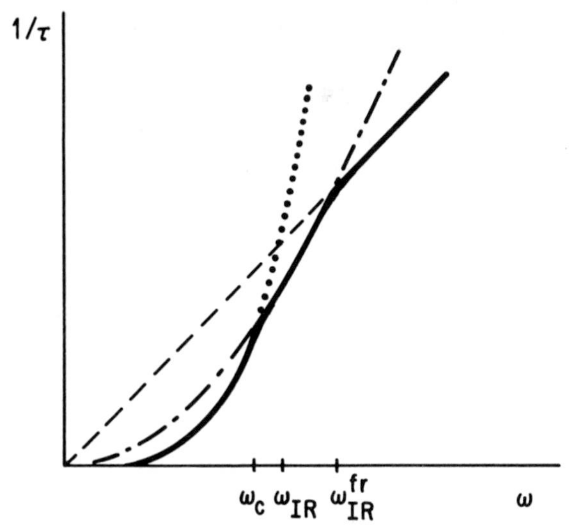

Fig. 3. Schematic plot of the vibration lifetime according to Eqs. (4), (20), and (21) (solid line). The dotted line is Rayleigh law, Eq. (4), and the dash-dotted line is the fracton lifetime, Eq. (20). The dashed line is the frequency.

We thus conclude that the "quantum" nature of the fractons, as exhibited by their lifetime [Eqs. (19) and (20)], together with the Ioffe-Regel condition, implies the existence of a "quantum fracton dimensionality" $\bar{\bar{d}}_q = 4/3$. For systems for which $\bar{\bar{d}} = \bar{\bar{d}}_q = 4/3$, the width $1/\tau$ obeys Eqs. (12) and the fractons are always in the Ioffe-Regel limit. In systems for which $\bar{\bar{d}} < 4/3$, scaling breaks down and two characteristic frequencies, at least, are required to specify the vibrational spectrum. In this case, if the fracton Ioffe-Regel frequency ω_{IR}^{fr} is smaller than the upper frequency cutoff of the spectrum (corresponding to the zone boundary), the vibrations at frequencies above ω_{IR}^{fr} will be in the Ioffe-Regel limit, and will be described by the dimensionality $\bar{\bar{d}}_q = 4/3$.

5. Conclusions

It is of interest to use our results for the phonon regime to consider the diffusion coefficient of the phonons. The electronic ("classical") diffusion coefficient of random materials scales as $\xi^{-\theta}$ at length scales longer than ξ [17]. For the phonons we obtain a different scaling. The square of the velocity of sound indeed scales like $\xi^{-\theta}$ [Eq. (6)], but multiplying it by the mean free time τ to obtain the diffusion coefficient introduces an additional power of ξ. Using Eqs. (4), (5), and (21), the diffusion coefficient of the phonons is

$$D(\omega) \sim \omega^{-d-1} \xi^{-\theta-(D/\bar{\bar{d}})(d-4+3\bar{\bar{d}})} , \qquad \omega < \omega_c , \qquad \bar{\bar{d}} \le 4/3 . \tag{22}$$

The (negative) exponent of ξ here is considerably larger than θ. For $d = 3$, $\bar{\bar{d}} = 4/3$, and $D = 2.5$, one has 7.375 for the exponent of ξ in (22), while $\theta = 1.75$. This implies a much stronger decay of $D(\omega)$ with increasing ξ, as compared with the electronic ("classical") result.

In summary, we have reviewed the basic features of the fracton scaling model for amorphous systems and tried to explore the implications of their characterization by a single length scale in the context of scattering theory. Assuming scaling, our results show that for the fractons to be in the strong scattering regime, and to obey the Ioffe-Regel condition, $\bar{\bar{d}}$ must be 4/3. This forces the Ioffe-Regel fre-quency of the phonons, ω_{IR}, to be proportional to the crossover frequency ω_c.

An interesting conclusion of our considerations is the possibility to examine the assumption of a single length scale describing the fractons, by carrying out scattering measurements in the phonon regime. As the scaling conjecture implies that $\omega_{IR} \sim \omega_c$, it determines the scaling of Rayleigh law with the crossover frequency. In the Brillouin scattering experiments by Courtens et al. [19] it was indeed found that the curve of $(\tau\omega_c)^{-1}$ as a function of ω/ω_c is independent of the density (which determines ω_c in their samples). Thus, this experiment implies $\omega_{IR} \sim \omega_c$. An attempt to fit their data with two different characteristic frequencies, ω_c and ω_{IR}, using Eq. (21) to relate them, has yielded the values [26] $D = 2.39 \pm 0.01$, $\theta = 1.51 \pm 0.1$, and $\bar{\bar{d}} = 1.36 \pm 0.04$, in a very good agreement with the theoretical values for a three-dimensional percolating network, and the prediction $\bar{\bar{d}} = 4/3$, derived from scaling arguments.

Acknowledgments

I would like to thank S. Alexander, R. Orbach, and A. Aharony for many helpful discussions. This work was supported in part by the National Science Foundation, grant 84-12898, and by the fund for Basic Research administered by the Israel Academy of Sciences and Humanities.

References

† Permanent address: School of Physics and Astronomy, Tel Aviv University, Tel Aviv 69978, Israel.

1. A. F. Ioffe and A. R. Regel, Progress in Semiconductors 4, 237 (1960); N. F. Mott, Phil. Mag. 19, 835 (1969).
2. R. C. Zeller and R. O. Pohl, Phys. Rev. B 4, 2029 (1971).
3. J. E. Graebner, B. Golding, and J. C. Allen, preprint, 1986.
4. Lord Rayleigh, Theory of Sound (McMillan, London, 1896), Vol. II.
5. S. John, H. Sompolinsky, and M. J. Stephen, Phys. Rev. B 27, 5592 (1983).
6. E. Akkermans and R. Maynard, Phys. Rev. B 32, 7850 (1985).
7. P. W. Anderson in T. Holstein Memorial Volume, ed. R. Orbach (Los Angeles, 1986).
8. S. Alexander and R. Orbach, J. de Physique Lett. 43, L625 (1982).
9. S. Alexander, C. Laermans, R. Orbach, and H. M. Rosenberg, Phys. Rev. B 28, 4615 (1983).
10. A. Aharony, S. Alexander, O. Entin-Wohlman, and R. Orbach, Phys. Rev. B 31, 2565 (1985).
11. S. Alexander, O. Entin-Wohlman, and R. Orbach, J. de Physique Lett. 46, L549, L555 (1985); Phys. Rev. B 32, 6447 (1985); Phys. Rev. B 33, 3935 (1986).
12. S. Alexander, O. Entin-Wohlman, and R. Orbach, Phys. Rev. B 32, 8007 (1985).
13. S. Alexander, O. Entin-Wohlman, and R. Orbach, Phys. Rev. Lett. (submitted); Phys. Rev. B 34, 2726 (1986).
14. A. K. Raychaudhuri, Ph.D. Thesis, Cornell University (1980), unpublished; J. E. de Oliviera and H. M. Rosenberg, private communication (1986).
15. A. Aharony, S. Alexander, O. Entin-Wohlman, and R. Orbach, Phys. Rev. Lett. (submitted); S. Alexander, Proc. of Stat. Phys., ed. H. E. Stanley (Boston, 1986).
16. S. Alexander, Ann. Isr. Phys. Soc. 5, 144 (1983).
17. Y. Gefen, A. Aharony, and S. Alexander, Phys. Rev. Lett. 50, 77 (1983).
18. Although we use the language of "scalar" elasticity [S. Alexander, J. de Physique 45, 1939 (1984)], the results equally apply for bending forces [I. Webman and Y. Kantor, in Kinetics of Aggregation and Gelation, eds., F. Family and D. P. Landau (North Holland, Amsterdam, 1984)].

19. E. Courtens, J. Pelous, J. Phalippou, R. Vacher, and T. Woignier, preprint, 1986.

20. R. Rammal and G. Toulouse, J. de Physique Lett. 44, L13 (1983).

21. S. Kelham and H. M. Rosenberg, J. Phys. C14, 1737 (1981).

22. Y. J. Uemura and R. J. Birgeneau, preprint, 1986.

23. One notes that the average in Eq. (15) is taken in a fractal space of dimensionality D. The reason for it is that we have used for the phonon density of states, Eq. (10), the form normalized in this space. In the percolation language, this means that we have considered the infinite cluster alone. When the finite clusters are included as well, to ensure that the overall density is uniform, the prefactor in the phonon density of states is changed into $\omega_c^{-d\theta/(2+\theta)}$ (see Ref. 10). The averaging is then carried out in Euclidean space and our final expression for τ^{-1}, Eq. (15), is unchanged.

24. A. Aharony, Y. Gefen, and Y. Kantor, J. Stat. Phys. 36, 795 (1984); Scaling Phenomena in Disordered Systems, eds., R. Pynn and A. Skjeltorp (Plenum, New York, 1985).

25. A. B. Harris, S. Kim, and T. C. Lubensky, Phys. Rev. Lett. 53, 743; 54, 1088 (1984); D. C. Hong, S. Havlin, H. J. Herrmann, and H. E. Stanley, Phys. Rev. B 30, 4083 (1984); J. G. Zabolitzky, ibid., 4077; H. J. Herrmann, B. Derrida, and J. Vannimenus, ibid., 4080; R. Rammal, J. C. Angler d'Auriac, and A. Benoit, ibid., 4087; J. C. Lobb and D. J. Frank, ibid., 4090; F. Leyvraz and H. E. Stanley, Phys. Rev. Lett. 51, 2048 (1983); A. Aharony and D. Stauffer, Phys. Rev. Lett. 52, 2368 (1984).

26. E. Courtens, private communication.

Condensed Matter Physics: The Theodore D. Holstein Symposium

Philip W. Anderson

It is a pleasure to write an appreciation of Ted Holstein, because he affected my own career so profoundly and so positively and because his legendary modesty always left one free to say all the appropriately positive things without any fears that he, wherever he is, will ever take them seriously.

The first time I encountered Ted was when I was searching for a job during the brief but sharp '49 depression, and visited Westinghouse. My thesis - I do not match Ted in modesty - has since become something of a classic, but at the time, even my thesis professor was not quite sure it was either correct or interesting. Thus it was that from Ted Holstein I first received that most satisfying of all professional compliments - real understanding of what I had done, and acceptance that it was genuinely original and might even be useful. My work was in line-broadening in gasses, and he was in the midst of doing a tour-de-force series of papers on propagation of resonance radiation in mercury; but they had in common the style, that was very new at that time, that considered the quantum many-body behavior of such a system as a gas containing an infinite number of molecules to be an approachable problem. He was almost the only person I met who was speaking the same language.

I did not go to Westinghouse in spite of Ted - primarily because he was too honest to promise me that I would work with <u>him</u>, not a much inferior group concerned with transistors - but he had left a very strong impression. When I encountered his name again in the Holstein-Primakoff theory of spin waves, which was one of the major inspirations for my own work on spin waves and other collective excitations, it was again as very early and unrecognized pioneer of the whole field of quantum theory of many-body systems.

Over the years it was always a pleasure to meet and talk to Ted - a pleasure which was all too rare because you had to go to him, since he never came to meetings. When he was at last elected to the NAS about 1975, he broke the habit of non-travel for once, and came to Washington to "sign the book". He stood - or sat - for about one day of meetings and quit. I know that for a fact because my wife came back from a day at the Ladies' program and told me what a delightful - and gossipy - companion he had been sitting on the bus together.

He was a superb physicist and a fine human being.

My First Meeting with Ted Holstein

Marvin L. Cohen

I first met Ted Holstein in 1963 when I was being interviewed by Westinghouse for a position in their research laboratories. The meeting took place during my seminar describing my PhD. thesis research predicting superconductivity in degenerate semiconductors. This was one of my first talks on the subject, and I was hoping to get Westinghouse experimentalists interested in the predictions (John Hulm did and collaborated in the experimental discovery). I was also anxious to get a job offer so that I could do theoretical work on a variety of experimental data on SiC obtained by Jim Choyke and Lyle Patrick.

The introductory part of the talk went smoothly, but it wasn't long before Ted began interrupting with questions, and the rest of the talk was almost a collaborative endeavor with Ted. He began with general probes asked in a tone which made me feel that he didn't believe anything I had said or was about to say. Soon the questions became more detailed. I recall him asking about how I could use a random phase approximation for a doped semiconductor since the density of carriers is low. I explained that r_s was still small because the relevant Bohr radius is large. He seemed to be satisfied with my electron-electron discussion, and I felt a little better about the likelihood that my future plan of working with Choyke and Patrick would come to pass. The next round of questions dealt with electron-phonon couplings, and the depth of the questions surprised me.

It seemed like every answer I gave resulted in two new questions. Ted proceeded along a path through my calculations starting from the foundations; he ventured down the alleys, turns, and dead-ends that I thought were my intellectual property alone. I remember thinking, "How could he know all this stuff? He hasn't seen my calculation." What started out as an uncomfortable inquisition was now a pleasurable joint investigation and a discussion about all those little details that mattered so much to me. He was testing for leaks in the logic and computations, and I was grateful. It was comforting to know that I was entering a field where people communicated on this level.

After the seminar and a few more questions from Ted, I managed to get Jim Choyke's ear and asked, "Who is that guy?" Jim said, "Ted Holstein, and you passed!" It wasn't until I returned to the University of Chicago that I learned the details about Ted's excellent reputation and his many important contributions.

I didn't see Ted again until much later. We first met Beverlee at LT16, the Holsteins visited Berkeley, and we spent time together at annual meetings in Washington. I remember discussions with Ted while we waited for late afternoon planes to the West Coast. He told me of his worries about some of the young theorists he'd heard at seminars. He questioned whether they really knew the physics behind all the formalism they presented because he wasn't always satisfied with the answers they gave to his questions. I remember wondering whether Ted was really satisifed with my answers 20 years earlier, but fortunately the subject never came up.

The Hall Effect Saga

by Lionel Friedman

In addition to papers of a technical nature, another intent of this Festshrift is to present reminiscences of a personal nature which convey to the reader the kind of physicist and person that Theodore Holstein was. As his first student, I was involved in a problem that at the time was among his favorites: the Hall mobility of the small polaron.

The two classic papers of 1959, "Studies of Small Polaron Motion," and II, (1) had just been published. Though there were earlier papers on the small polaron problem in the Soviet (2) and Japanese (3) literature, the latter in particular treated only the hopping regime. The second Annals of Physics paper was the first to treat the polaron band motion and hopping motion on an equal footing as due to the diagonal and off-diagonal matrix elements of the intersite transfer, respectively. One possible application of this theory was to the transition metal oxides (e.g., NiO, MnO), which were then of interest at the Westinghouse Research Laboratories as possible thermoelectric materials. The Hall effect in the hopping regime was then an outstanding question. Since the Lorentz force did not act coherently on extended state motion, many said that a Hall effect did not exist here; other arguments of a macroscopic nature said that it should. A microscopic theory did not exist.

Before starting my work, I was told to expect a stern supervisor. However, though I found him totally committed and demanding, he was at the same time very kind and understanding of me as a student. My initial investigations treated the magnetic field as a perturbation. Also, it soon became clear that the magnetic field could not alter the lowest order two-site jump rate. At that time, Ted Holstein found a paper by G.E. ZILBERMAN (4) utilizing the Peierls phase factors in a study of magnetism in tight-binding approximation. The incorporation of these phase factors in the molecular-crystal-model was then straightforward. The well-known picture of the Hall effect as arising from higher order jump processes involving one (or more) intermediate sites followed immediately. (5) Ted called this a "microscopic Bohm-Aharanov effect." It remained to develop perturbation theory to third order in the intersite transfer for the fully quantum mechanical and classical regimes. Indeed, it was possible to obtain the same result from a number of different approaches.

Now the above microscopic mechanism for the Hall effect was applicable to any hopping mechanism. In particular, the experimental observation of the Hall effect appeared most promising for the a.c. Hall effect in single phonon-assisted impurity conduction. However, before he began to develop the theory for this case, Ted was concerned that his work would appear in print before my small polaron work (as it did).

I was then told that, in the event that this should happen, the first sentence of his impurity conduction paper would acknowledge the earlier small polaron studies. Indeed, such a sentence does appear in that paper. (6)

In later years, our work was criticized by German (7) and Russian (8) workers on the grounds that it used a jump probabilistic formalism rather than a linear response approach starting from the Kubo formula. Though he never avoided a formal approach, the physics always took precedence. The Russian work, in fact, ascribed a negative activation energy to the Hall mobility, i.e., the absence of potential barriers for the transverse motion, a manifestly unphysical result. Ted Holstein was not one to allow a controversy in the theory to remain unresolved. This motivated a derivation of the off-diagonal conductivity from Kubo's formula which was in complete agreement with our earlier result. (9) The Russian result appeared to have been due to a calculational error.

Another quality of Ted Holstein's was his insistence on a high level of exposition. I rewrote the introduction to my thesis seven or eight times until it met his high standards, and I consider my writing the better for it.

Ted recognized the relevance of the now classic paper of KOHN and LUTTINGER (10) on the derivation of the Boltzmann transport equation from the density matrix formalism, based on the separation of its diagonal and off-diagonal matrix elements. Indeed, he had me give a series of talks on this paper. Such an approach was able to explain small polaron mobility as the additive contributors (11) of the band and hopping terms in a natural and straightforward way.

Further developments in the Hall effect were the following: first, the interference mechanism for the Hall effect in hopping was applied to the polaron-band regime where it was shown that the Peieris phase factors yield the classical Lorentz force in a Boltmann equation description (12) of small polaron motion. In this latter regime, the sign anomalies for the hopping case could more physically be understood as arising from the appropriate weighting of the negative mass states in the narrow band limit (bandwidths $<\sim k_B T$). The same model proved useful in explaining transport properties of organic molecular crystals (13) where the bands are narrow, but where self-trapping plays no role.

The Hall effect was extended by EMIN and HOLSTEIN to the four-site case (14) and to the adiabatic (large J) regime, respectively, (15) where the microscopic mechanism is different from that described earlier. In this connection, Ted was gracious in acknowledging (15) his interaction with Dr. Conyers Herring, who suggested this mechanism. Small polaron theory and the Hall effect have found application to a number of material systems. The three-site mechanism also served as a model for the Hall effect in amorphous and liquid semiconductors.

The three-site mechanism for the Hall effect was applied to the "diffusive regime" for the just-extended states at the mobility edge. (16) However, the charge transport mechanism here is much less well defined than for the self-trapped or extended-state limits. Though the suggestion that small polarons are the principal charge carriers in amorphous solids remains controversial (17), it must be said that there is at present no good alternative theory of the Hall effect sign anomalies (especially) in these materials. Ted Holstein felt that the necessity of having to know detailed microscopic information (e.g., site coordination, local wavefunctions, etc.), in order to be able to predict the Hall effect, made the latter a less useful theoretical tool. Also, it must be said that the inability to measure an a.c. Hall effect in impurity conduction is distressing (18); it is hoped that this will be observed at some future time.

References

1. T. Holstein, Ann. Phys. (N.Y.) 8, 325, 343 (1959).
2. S.V. Tyablikov, Zh. Eksperim i Teor. Fa. 23, 381 (1952).
3. J. Yamashita and T. Korosawa, Phys. Chem. Solids 5, 34 (1958).
4. G. Zilberman, Soviet Phys. - J.E.T.P. 2, 650 (1956).
5. L. Friedman and T. Holstein, Ann. Phys. (N.Y.) 21, 494 (1963).
6. T. Holstein, Phys. Rev. 124, 1329 (1961).
7. J. Schnakenberg, Z. Physik 185, 123 (1965).
8. Yu A. Firsov, Fiz. Tverd. Tela 5, 2149 (1963) [English transl.: Soviet Phys. - Solid State 5, 1566 (1964)].
9. T. Holstein and Lionel Friedman, Phys. Rev. 165, No. 3, 1019 (1968).

10. W. Kohn and J.M. Luttinger, Phys. Rev. 108, 590 (1957).
11. L. Friedman, Phys. Rev. 135, A233 (1964).
12. L. Friedman, Phys. Rev. 131, 2445 (1963).
13. L. Friedman, Phys. Rev. 133, A1668 (1964).
14. D. Emin, Ann. Phys. (N.Y.) 64, 336 (1971).
15. D. Emin and T. Holstein, Ann. Phys. (N.Y.) 53 439 (1969).
16. L. Friedman, J. Non-Crystalline Solids, 6, 329 (1971).
17. C.H. Seager, D. Emin and R.K. Quinn, Phys. Rev. B 8, 4746, (1973), and later papers.
18. Robert Klein, Thesis, Univ. of California at Riverside and Physical Review B 31, 2014 (1985).

REMEMBRANCES OF TED HOLSTEIN

Edward Gerjuoy

I first became acquainted with Ted Holstein in 1937. I had just graduated from CCNY, now the City College of CUNY, but had not yet gone on to graduate school. I had developed the habit of quietly auditing--on a tuition unpaid basis of course--several of the graduate courses offered by the NYU physics department. Ted was one of the graduate students regularly enrolled in these courses, and it took me no time at all--listening from my vantage point at the back of the room where I never dared ask a question for fear of being thrown out--to determine that he was one of the very few students who understood what was going on. So after class I used to buttonhole Ted and ask him to explain what I hadn't understood. Ted, already set in the behavior patterns which continued throughout his life, always answered my questions with great clarity, and with insight into the physical significance of the mathematical results. I would not say that he never was impatient, but I never felt he was putting me down, and he seemed willing to spend astonishing amounts of time to get across a point he felt was important.

As it turned out, that acquaintance with Ted which started in 1937 had a predominant, perhaps determinative influence on the course of my own career. I maintained casual contacts with him through my own graduate student years and during World War II, and thus knew that Ted had received his Ph. D. from Otto Halpern, whom he regarded as a very good physicist though rather "difficult" to work with. Therefore, when I was at USC after the war, and Halpern was suggested for a senior professorship, I strongly supported his appointment, figuring that Ted's "very good physicist" recommendation was highly reliable, but placing less reliance on his assessment of Halpern's personality. Later, when Halpern had come to USC, I realized that I should have given credence to Ted's judgment on all counts. In any event, Halpern's tenure at USC soon gave rise to such turmoil in the physics department that I jumped at an offer to spend the year 1951-52 visiting at the NYU physics department. At the New York meeting that year, I had a chat with Ted, who suggested that I spend the summer at Westinghouse Research Labs in Pittsburgh on my way back West from NYU. I accepted that offer, and during the summer gave a talk at the University of Pittsburgh physics department which resulted in a tenured position offer, which I also accepted. I remained a consultant at Westinghouse, in Ted's group, while teaching at Pitt, and it was a suggestion by Ted, offered unselfishly and with his customary insight, that set me to calculating atomic collision cross sections (in this case the cross sections for rotational excitation of molecular nitrogen by slow electrons), the theoretical physics area on which I concentrated for many many years. Thus it is accurate to say that were it not for Ted Holstein, I almost certainly would not now be at the University of Pittsburgh, and would not have turned to atomic collision theory from the nuclear physics problems on which I previously had concentrated.

As for anecdotes about Ted, they are legion, as is to be expected for a person with so strong and unusual a personality. I recall Ted being almost tearfully requested by Ed Creutz, then the chairman of the Carnegie Tech (now Carnegie-Mellon University) physics department, to either refrain from asking questions at seminars and colloquia or else not to attend, this after Ted had delivered a particularly severe battering to a poorly prepared colloquium speaker. I recall many happy hours discussing our mutual imaginary disease symptons with Ted, whose hypochondria was an able match for my own. His hypochondria led him to focus on physical fitness, however, something I was too lazy to copy. In this connection one last anecdote, which captures most of the essence of Ted's personality, may be worth recounting. Ted was in the habit of going to the Y and having a swim before coming to work. One morning I was at Westinghouse, as part of my consulting, when Ted walked in, or rather dragged himself in--he seemed barely able to stand up. "Ted," I asked, with real concern, "what's the matter? Are you sick?" "No, I'm OK. It's just that I decided to swim seventy five laps, fast. Toward the end I didn't think I'd make it."

A Few Remembrances of Ted Holstein

Bernd Schüttler

I'd like to begin by thanking everyone–the organizers, the participants, and the sponsors–who made this symposium possible. It's been a great tribute to Ted.

I have been asked to make a few remarks tonight regarding Ted Holstein. I should preface them by mentioning that I was one of Ted's last Ph.D. students and spent about three years working together with him at UCLA. Thinking about the many discussions that we had, one conversation comes to my mind which probably demonstrates very nicely Ted's approach to theoretical physics and the standards which he applied to his research:

We had been working for some time on a problem concerning one-dimensional polaron transport. Our project had progressed to a point where we had obtained the most important results, and, I thought, we basically understood the physics of our model. So, one day, I suggested to him that our results might be useful to explain some recent experimental data. Shouldn't we make more of an effort to apply our theory to the experiments (rather than working out all these theoretical details out of which his famous "appendices" are made)? It was then that Ted became very quiet and, with a rather sad look in his face, he finally responded, "I really didn't know that <u>that</u> was your idea about doing theoretical physics. Right now, I can't be bothered by the experiments. It is first of all our job, to make an honest theory. That's the only way that we will be able to tell <u>them</u> (the experimentalists) something instead of them telling us."

It was Ted's approach, to study each aspect of a problem very carefully until each little facet of it was completely understood—if a certain point remained unclear, the paper needed another appendix. Those who have worked with him or studied his work, will appreciate the length to which he carried this principle. Along this line, I should mention that our last paper together has recently appeared in print–our paper <u>and</u> its seven appendices!

Thank you for the opportunity to express my thoughts regarding Ted Holstein and thank you again for attending this memorable meeting.

Remembrance of Theodore D. Holstein

Lawrence A. Vredevoe

Ted Holstein joined the faculty while I was a graduate student in physics at UCLA. His graduate courses in condensed matter physics proved invaluable to me in preparation for my doctoral research on phonon interactions in solids. It was an honor to have him sit on my doctoral committee.

Ted displayed a depth of insight and needed skepticism that was so important to the young graduate student all too eager to quickly reach conclusions. His help with difficult theoretical problems will always be remembered.

Although my career interests eventually led me from physics into medicine, the analytical thought process nurtured in me by Ted and other outstanding physicists has proved to be of lasting benefit. Four years ago I helped care for Ted as a physician, and was honored to be able in some small way to reciprocate for all the care he had shown for me.

Some Background to the Accomplishments of Ted Holstein

A. Theodore Forrester

Any physicist who has attended seminars at which Ted Holstein was in attendance is aware of his insistence on a total understanding and an ability to focus on the key questions, questions which illuminated the subject for all. Edward Gerjuoy, a fellow attendee with him at graduate courses, related that even at that stage in Ted's career it was clear that he was one of those few students who could be counted on to understand what was going on and was a great asset in his willingness and ability to communicate clearly his understanding.

While most regarded Ted as a great asset at seminars, this view was not universal. For example, there was an incident at Carnegie Tech (now Carnegie-Mellon University) in which Ted delivered a particularly severe battering to a poorly prepared colloquium speaker. The Chairman of the Physics Department was so embarrassed for his guest that he almost tearfully requested Holstein to either refrain from asking questions or not to attend the Carnegie Tech colloquia.

The meticulousness with which Ted approached the question of understanding carried over to the thoroughness and completeness of his own contributions and were, naturally enough, reflections of Ted's personality generally. Ted Forrester, for example, recalls squatting besides Holstein on a lawn in Pittsburgh as Holstein dug up a small quantity of weeds whose number must surely have totalled in the millions. One must approach this, he stated, with the patience of a Chinese philosopher by which it seems he meant that imperfections, where recognized, were not to be tolerated, even if one cannot possibly get them all.

The rigidity of his exercise regimen is another example of Ted's compulsiveness. He was encountered, on one occasion, barely able to stand. "Are you ill?" he was asked. "No, I'm OK. It's just that I decided to swim seventy five laps, fast. Toward the end I didn't think I'd make it." This sort of determination, as well as his brillance, has to be responsible for his extraordinary contributions to which area he turned his attention.